IC DESIGN PROJECTS

By Stephen Kamichik

IC DESIGN PROJECTS

By Stephen Kamichik

©1998 by Howard W. Sams & Company

PROMPT© Publications is an imprint of Howard W. Sams & Company, A Bell Atlantic Company, 2647 Waterfront Parkway, E. Dr., Indianapolis, IN 46214-2041.

All rights reserved. No part of this book shall be reproduced, stored in a retrieval system, or transmitted by any means, electronic, mechanical, photocopying, recording, or otherwise, without written permission from the publisher. No patent liability is assumed with respect to the use of the information contained herein. While every precaution has been taken in the preparation of this book, the author, the publisher or seller assumes no responsibility for errors or omissions. Neither is any liability assumed for damages resulting from the use of information contained herein.

International Standard Book Number: 0-7906-1135-X

Acquisitions Editor: Candace M. Hall
Editor: Natalie F. Harris
Assistant Editors: Pat Brady, Loretta Yates
Editorial Support: Karen Mittelstadt, Shey Query
Typesetting: Natalie Harris
Indexing: Natalie Harris
Cover Design: Christy Pierce
Illustrations: Stephen Kamichik, Bill Skinner

Trademark Acknowledgments:
All product illustrations, product names and logos are trademarks of their respective manufacturers. All terms in this book that are known or suspected to be trademarks or services have been appropriately capitalized. PROMPT© Publications, Howard W. Sams & Company, and Bell Atlantic cannot attest to the accuracy of this information. Use of an illustration, term or logo in this book should not be regarded as affecting the validity of any trademark or service mark.

PRINTED IN THE UNITED STATES OF AMERICA

9 8 7 6 5 4 3 2 1

◆ Contents ◆

Preface   1

PART ONE: POWER SUPPLIES   3

Chapter 1
Power Supplies   5
   Transformers   6
   Rectifiers   7
      Half-Wave Rectifier   8
      Full-Wave Rectifier   8
      Full-Wave Bridge Rectifier   10
   Regulators   12
      Zener Diode Shunt Regulator   14
      Transistor Series Regulator   16
   Computer-Aided Power Supply Component Selection   17
   Problems   22

Chapter 2
IC Regulators   27
   Linear IC Regulator   27
   Switch-Mode Integrated Circuit Regulator   28
   MC7800 and MC7900 Regulators   32
   LM723 IC Voltage Regulator   36
   LM317 and LM337 Adjustable Regulators   43
   Problems   48

Chapter 3
Battery Charger   49
   Circuit Operation   49
   Construction   51
   Testing and Use   52

Chapter 4
Bipolar Power Supply — **53**
- Circuit Description — 54
- Construction — 54
- Testing and Use — 55

Chapter 5
5-Volt Power Supply — **57**
- Circuit Description — 57
- Example 5-1 — 60
- Construction — 60
- Testing and Calibration — 61
- Using the Power Supply — 63

Chapter 6
Dual-Tracking Power Supply — **67**
- Circuit Description — 67
- Construction and Use — 68
- Circuit Modifications — 68

PART TWO: TTL & CMOS LOGIC FAMILIES — 71

Chapter 7
Transistor-Transistor Logic (TTL) — **73**
- Input Circuit Operation — 74
- Output Circuit Operation — 74
- Circuit Operation with Input High — 75
- Circuit Operation with Input Low — 76
- TTL NAND Gate — 77
- TTL Specifications — 78
- Other Forms of TTL Gates — 79
- Problems — 80

Chapter 8
CMOS Logic — **83**
- MOS Transistors — 83
- CMOS Inverter — 84
- CMOS NOR and NAND Gates — 86

Specifications	86
Proper Handling of CMOS Integrated Circuits	88
Problems	89

Chapter 9
Capacitance Meter — 91

Circuit Description	91
Construction and Verification	94
Calibration and Use	95

Chapter 10
Digital Logic Probe — 97

Circuit Description	98
Construction	101
Testing and Use	102

PART THREE: OPERATIONAL AMPLIFIERS — 105

Chapter 11
The Operational Amplifier — 107

The Ideal Operational Amplifier	107
The Practical Operational Amplifier	109
Specifications	110
Operational Amplifier Circuits	113
Linear Circuits	113
Nonlinear Circuits	115
Problems	116

Chapter 12
Home Theater System — 119

Circuit Operation	121
Construction	125
Testing	126
Installation	126

Chapter 13
Alpha Brain-Wave Feedback Monitor — 127

The Brain as a Voltage Source	129

Biofeedback	130
Circuit Operation	131
Construction	133
Electrodes	133
Balancing the Instrumentation Amplifier	134
Using the Alpha Brain-Wave Feedback Monitor	135
Possible Improvements	136

Chapter 14
IC Stereo Preamplifier — 137

Circuit Description	137
Construction	139

PART FOUR: PHASE-LOCKED LOOP — 141

Chapter 15
Phase-Locked Loop — 143

PLL Commercial Applications	144
LM565 PLL Integrated Circuit	145
Applications for the LM565	147
CMOS PLL Integrated Circuit	148
Problems	151

Chapter 16
Function Generator — 153

Circuit Operation	153
Construction	154
Testing and Use	155

Chapter 17
Siren — 157

Circuit Operation	157
Construction	159

Chapter 18
Audible Logic Probe — 161

Circuit Operation	161
Construction	163
Testing and Use	164

PART FIVE: LM555 TIMER IC — 165

Chapter 19
The LM555 Timer IC — 167
- Inside the LM555 — 167
- Astable Operation — 169
- Monostable Operation — 171
- Applications — 173
- Problems — 177

Chapter 20
DC Motor Speed Controller — 181
- Circuit Operation — 181
- Construction — 183
- Other Uses — 184

Chapter 21
Electronic Organs — 185
- Circuit Description — 187
- Construction — 189
- Tuning and Use — 192

Chapter 22
Automatic Light Timer — 195
- Circuit Description — 195
- Construction — 198
- Calibration and Use — 199

Chapter 23
Warble Alarm — 201
- Circuit Description — 201
- Construction — 203
- Test Procedure — 204

Chapter 24
Darkroom Timer — 205
- Circuit Operation — 205

Construction	207
Testing and Use	208

PART SIX: LM567 TONE DECODER IC — 209

Chapter 25
LM567 Tone Decoder IC — 211
Applications	213
Problems	215

Chapter 26
Two-Channel Infrared Remote Control — 217
Transmitter Circuit Operation	217
Receiver Circuit Operation	219
Construction	223
Setup and Use	223

Appendix
Problem Solutions — 227
Chapter 1	227
Chapter 2	228
Chapter 7	229
Chapter 8	230
Chapter 11	231
Chapter 15	233
Chapter 19	234
Chapter 25	235

Glossary — 237

Bibliography — 241

Index — 243

◆ Preface ◆

There are literally thousands of integrated circuits (ICs) available. Most are specialized devices, and a handful are truly versatile integrated circuits. The purpose of *IC Design Projects* is to discuss some of the most popular and versatile integrated circuits and to offer you some projects in which you can learn to create useful and interesting devices with these ICs.

This book is divided into six parts. Part 1 discusses power supplies, and contains six chapters. Chapter 1 tells how power supplies operate. Chapter 2 deals with several integrated circuit voltage regulators. Chapter 3 describes a battery charger project. Chapter 4 details the construction of a bipolar power supply. Chapter 5 deals with a five-volt power supply project. Chapter 6 discusses the construction of a dual tracking power supply.

Part 2 discusses TTL and CMOS logic families, and contains four chapters. Chapter 7 is about TTL logic. Chapter 8 is about CMOS logic. Chapter 9 discusses a capacitance meter project. Chapter 10 discusses a digital logic probe project.

Part 3 is about operational amplifiers, and contains four chapters. Chapter 11 discusses how operational amplifiers work. Chapter 12 describes the construction of a surround-sound decoder. Chapter 13 discusses a brain wave monitor project. Chapter 14 details the construction of a stereo preamplifier.

Part 4 is about phase-locked loops, and contains four chapters. Chapter 15 discusses how phase-locked loops work. It also covers the LM565 and the CD4046 PLL integrated circuits. Chapter 16 details the construction of a function generator. Chapter 17 discusses a siren project. Chapter 18 describes the construction of an audible logic probe.

Part 5 is about the ubiquitous LM555 timer integrated circuit, and contains six chapters. Chapter 19 discusses the LM555 timer. Chapter 20 describes a DC motor speed control project. Chapter 21 details the construction of two electronic organ projects. Chapter 22 discusses the construction of an automatic light timer. Chapter 23 is about a warble alarm project. Chapter 24 discusses a darkroom timer project.

IC Design Projects

Part 6 is about the LM567 tone decoder integrated circuit, and contains two chapters. Chapter 25 discusses the LM567 tone decoder. Chapter 26 describes the construction of an infrared remote control system.

Part One
◆ Power Supplies ◆

Chapter 1
◆ Power Supplies ◆

Most electronic circuits require DC (direct current) voltage sources or power supplies. If the electronic device is to be portable, then one or more batteries is usually needed to provide the DC voltage(s) required by the electronic circuits.

Electronic circuits require a stable DC voltage source for a stable Q or quiescent operating point. Batteries generate a pure DC voltage. Batteries have a limited life span, and most cannot be recharged. They are also relatively expensive.

The solution for non-portable equipment is to convert the alternating current (AC) household line voltage to a DC voltage source. *Figure 1-1* is a block diagram of an AC-to-DC power supply. The transformer steps the household line voltage up or down, as required. The rectifier converts the AC voltage to a pulsating DC voltage. A regulator is used to smooth the pulsating DC voltage to a varying DC voltage. The better the regulator, the closer the carrying DC voltage approximates a pure or unvarying DC voltage.

The different types of DC voltages are shown in *Figure 1-2*. A pulsating DC voltage varies from zero volts to a maximum voltage. A varying DC voltage does not go to zero volts. The AC component is zero volts in a pure DC voltage source.

In North America, the power distribution frequency is 60 Hz. In Europe, it is 50 Hz. In aircraft, the power distribution frequency is 400 Hz, in order to reduce the size and weight of the power distribution equipment.

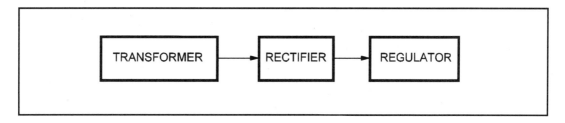

Figure 1-1. Block diagram of an AC-to-DC power supply.

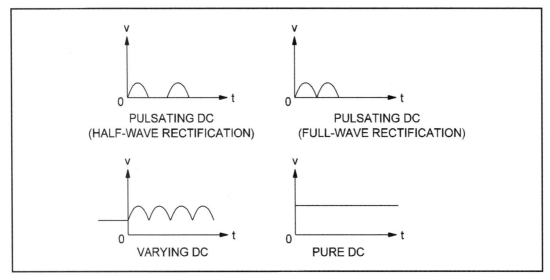

Figure 1-2. Different types of DC voltages.

Transformers

The basic transformer consists of a primary winding of N1 turns and a secondary winding of N2 turns, as shown in *Figure 1-3*. The primary and secondary windings are wound on a magnetic core. When an alternating line voltage, V1, is applied to the primary winding, an alternating current, I1, creates a flux. Part of the flux flows through the magnetic core, inducing an alternating current, I2, in the secondary winding; which in turn generates an alternating voltage, V2, in the secondary winding. When a load is placed across the secondary terminals, the alternating voltage, V2, produces an alternating current, I2, in the load. Placing a load on the secondary winding increases the primary current. The primary voltage remains fairly constant.

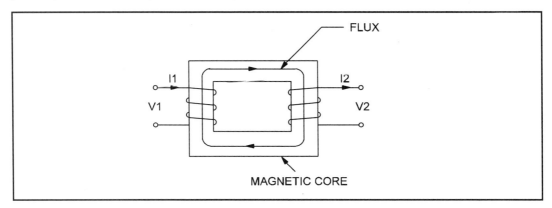

Figure 1-3. A basic transformer.

Power Supplies

Three important transformer specifications are the secondary voltage, the power rating and the regulation factor. The secondary voltage is specified in RMS volts at the full transformer-rated power load. The power load is specified in VAs (volt-amperes) or watts. A 20 VA transformer delivers 2 amps when its secondary voltage is 10 volts. The regulation factor is the percentage by which the transformer secondary voltage increases when the load is removed. A transformer with a 15-volt secondary at full load and a 16.5-volt secondary at no load is said to have a 10 percent regulation factor. The regulation factor is Reg = [V2(NL) - V2(FL)]/V2(FL).

A step-down transformer has a secondary voltage that is lower than the primary voltage. A step-up transformer has a secondary voltage that is higher than the primary voltage. An isolation transformer has a secondary voltage that is equal to the primary voltage.

Rectifiers

Three basic rectifier circuits are the half-wave rectifier, the full-wave rectifier and the full-wave bridge rectifier. A rectifier circuit converts an alternating voltage into a pulsating DC voltage. This is accomplished by using one or more diodes because

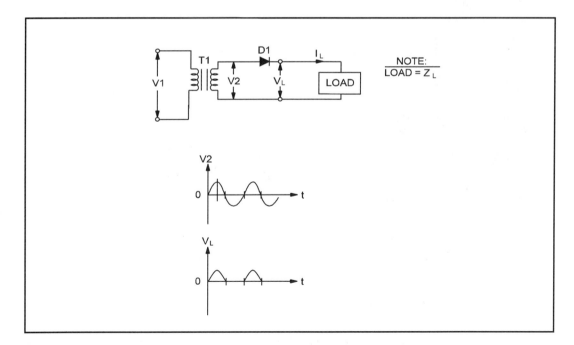

Figure 1-4. Half-wave rectifier circuit and waveforms.

diodes conduct current in only one direction. Diodes are discussed in more detail in *Semiconductor Essentials for Hobbyists, Technicians & Engineers*, also written by the author.

Half-Wave Rectifier

The simplest rectifier circuit is the half-wave rectifier, shown in *Figure 1-4*. The transformer isolates the household line voltage, and also steps down (or up) the household voltage to a more useful voltage level. The diode lets current flow into the load in only one direction. The load current is unidirectional; therefore, it has a significant DC component (or average value). The half-wave rectifier waveforms are also shown in *Figure 1-4*.

When V2 is positive, diode D1 conducts and V_L = V2. When V2 is negative, diode D1 blocks the current flow and V_L = 0 volts. The resulting load voltage has an average or DC voltage along with a considerable AC or ripple voltage.

Some useful design formulas are:

$V_{AVG}(DC) = 0.637 \, V2$

$V_L(DC) = 0.318 \, V2$

$V_L(AC) - 0.386 \, V2$

$\%Ripple = V_L(AC)/V_L(DC) = 121\%$

The ripple frequency of a half-wave rectifier circuit is the same frequency as the household line voltage. The diode must have a PRV (peak reverse voltage) rating equal to the maximum load voltage.

In a half-wave rectifier circuit, the ripple component is larger than the DC component, which is undesirable. For this reason, a full-wave rectifier circuit is preferred.

Full-Wave Rectifier

The full-wave rectifier circuit requires a transformer with two secondary windings; that is, a center-tapped secondary winding, as shown in *Figure 1-5*. The polarities of

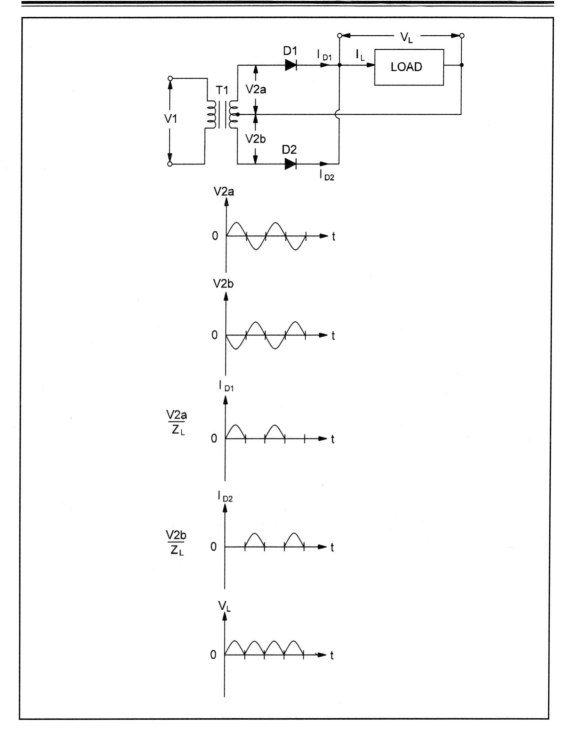

Figure 1-5. Full-wave rectifier circuit and waveforms.

the secondary windings must be 180 degrees out of phase. The diode waveforms are therefore 180 degrees out of phase. The load sums the two diode currents and produces the full-wave voltage. The full-wave rectifier waveforms are also shown in *Figure 1-5*.

Diode D1 conducts when V2a is positive, producing a half-wave rectified voltage across the load. Diode D1 does not conduct when V2a is negative. Diode D2 conducts when V2b is positive, producing a half-wave rectified voltage across the load. Diode D2 does not conduct when V2b is negative. One of the two diodes is conducting at all times because V2a and V2b are 180 degrees out of phase. The load takes the two half-wave rectified voltages and sums them, producing the full-wave rectifier voltage.

Some useful design formulas are:

$V_L(DC) = 0.637\ V2$

$V_L(AC) = 0.307\ V2$

$\%Ripple = V_L(AC)/V_L(DC) = 48.2\%$

The ripple frequency of a full-wave rectifier circuit is twice the frequency of the household line voltage. The PRV rating of the diodes must be equal to twice the maximum load voltage.

The full-wave rectifier circuit has a lower ripple factor than the half-wave rectifier circuit. In the full-wave rectifier circuit, the DC component of the load voltage is increased, and the AC component of the load voltage is decreased.

If a center-tapped transformer is not available, a full-wave bridge rectifier circuit may be used.

Full-Wave Bridge Rectifier

The full-wave bridge rectifier circuit requires four diodes, as shown in *Figure 1-6*. The diodes are arranged in a bridge configuration, hence the name. The transformer only has one secondary winding. The transformer utilization in the full-wave bridge rectifier circuit is improved over that of the full-wave rectifier circuit. The waveforms of the full-wave bridge rectifier are also shown in *Figure 1-6*.

Power Supplies

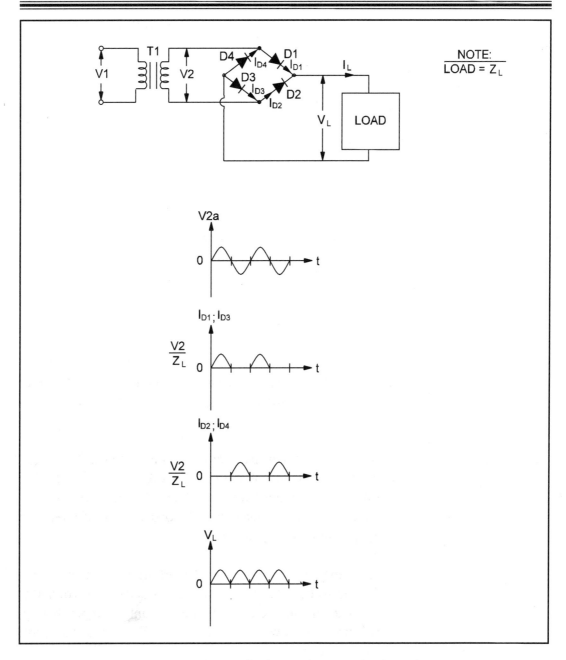

Figure 1-6. Full-wave bridge rectifier circuit and waveforms.

When V2 is positive, diodes D1 and D3 conduct current through the load. Diodes D2 and D4 block current flow when V2 is positive. When V2 is negative, diodes D2 and D4 conduct current through the load. Diodes D3 and D4 block current flow when V2 is negative. The load sums the two voltages and produces the full-wave bridge load voltage.

Some useful design formulas are:

$V_L(DC) = 0.637\ V2$

$V_L(AC) = 0.307\ V2$

% Ripple = $V_L(AC)/V_L(DC)$ = 48.2 %

The ripple frequency of a full-wave bridge rectifier is twice that of the household line voltage. The diode PRV rating must equal the maximum load voltage. The full-wave bridge rectifier fully utilizes the transformer secondary winding during both half cycles.

Regulators

Electronic circuits usually require a power supply that has a ripple factor of much less than 48.2%. A low-pass filter can be used to pass the DC (the frequency of DC is zero) component of the power supply voltage while reducing its AC component as much as possible.

The simplest regulator is a large capacitor in parallel (or shunt) with the load. The capacitor stores DC voltage while the load voltage increases to it peak value. The capacitor discharges or dumps its stored voltage while the load voltage decreases to its minimum value. The capacitor converts the pulsating DC load voltage of a rectifier into a smooth DC load voltage, as shown in *Figure 1-7*.

Two important parameters of a capacitor regulator are its working voltage and its capacitance. The working voltage must be at least equal to the no-load output voltage of the power supply circuit. The capacitance determines the amount of ripple that appears on the DC output when current is drawn from the circuit. The amount of ripple decreases when the capacitance increases.

A useful formula is $dQ = CdV = I_L dt$, where dQ is the charge lost in Coulombs, C is the capacitance in Farads, dV is the peak-to-peak ripple voltage in volts, I_L is the load voltage in amperes, and *dt* is the discharge time.

Power Supplies

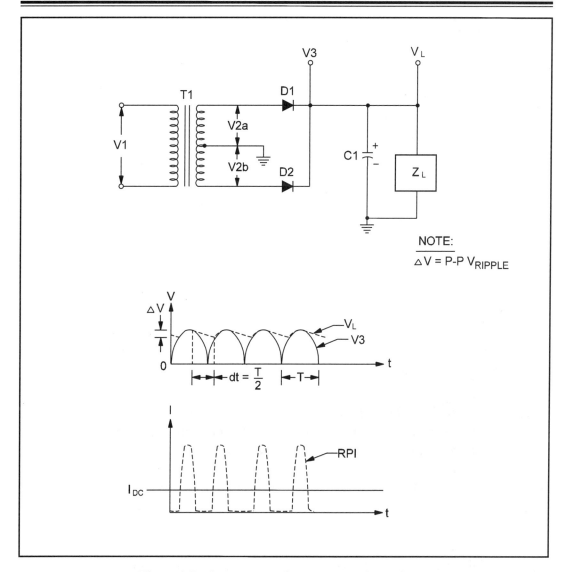

Figure 1-7. Capacitor regulator circuit and waveforms.

For half-wave rectifiers, dt = T/2 = 8.33 msec. For full-wave and full-wave bridge rectifiers, dt = T/2 = 4.167 msec.

The use of a capacitor regulator increases the repetitive peak (RPI) current flow through the rectifier diode(s). The RPI current flow increases as the capacitance increases. If the capacitor is large enough, the RPI can be ten times the direct current flow through the load.

IC Design Projects

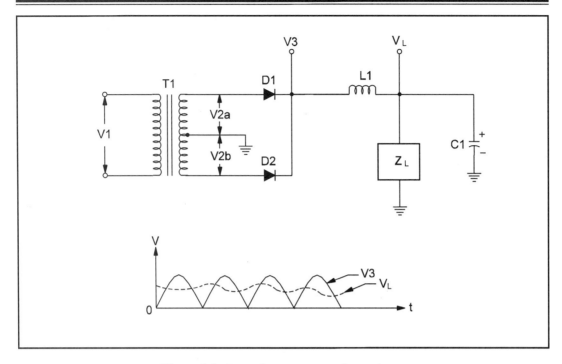

Figure 1-8. L-regulator circuit and waveforms.

The series choke (or inductor) regulator can also be used to reduce the ripple factor. The inductor stores energy during the peaks of the alternating current and releases it when the rectifier output falls below the load voltage.

The L-regulator consists of a series inductor and a capacitor in parallel with the load, as shown in *Figure 1-8*. The L-regulator is often used in high-power (1KW or more) DC supplies.

The Pi-regulator consists of two shunt capacitors and a series inductor, as shown in *Figure 1-9*. The Pi-regulator has a poor regulation factor as compared with the L-regulator. However, the Pi-regulator has a lower ripple factor because the input capacitor increases the filtering effect.

Zener Diode Shunt Regulator

A zener diode shunt regulator can be used to provide a maximum load current of 50 milliamperes. The basic zener diode regulator is shown in *Figure 1-10*. A current of about 5 mA flows through the zener diode, D1, from the power supply, V1, through the

Power Supplies

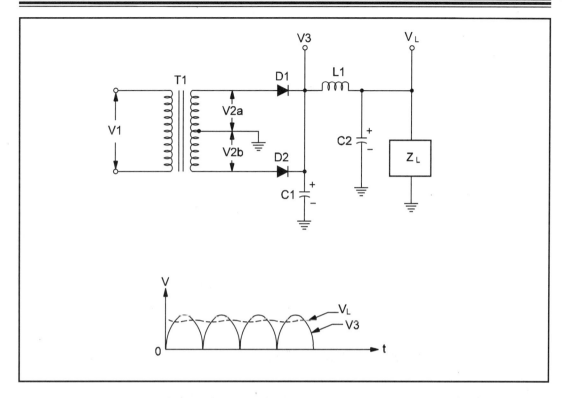

Figure 1-9. Pi-regulator circuit and waveforms.

current limiting resistor, R1. The load voltage, V2, is maintained at the zener voltage as long as V1 is a few volts greater than the zener voltage, Vz. The zener diode passes the full circuit current when no load current is drawn. When the full load current is drawn, only 5 mA of current flows through the zener diode.

Some useful formulas are:

$R1 = (V1 - V2)/(I_L + 0.005)$

$P_{R1} = (V1 - V2)(I_L + 0.005)$

$P_{D1}(\max) = V2(I_L + 0.005)$

The zener diode can be used as a reference voltage source in a transistor series regulator.

IC Design Projects

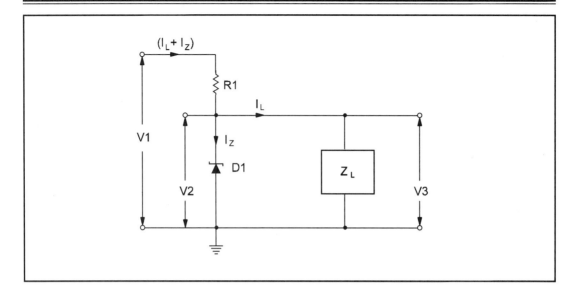

Figure 1-10. Zener diode shunt regulator.

Transistor Series Regulator

The transistor series regulator is shown in *Figure 1-11*. The zener diode, D1, provides a reference voltage to the base of transistor Q1. Transistor Q1 is configured as a voltage follower. The emitter voltage of Q1 is 0.6 volts below the base or zener diode reference voltage of Q1. Components D1 and R1 provide the base drive current to Q1. The base current is approximately equal to the load current divided by the current gain or beta of Q1. The higher the beta of the transistor, the better the output voltage regulation of the transistor series regulator.

The transistor passes the full load current. Resistor R1 passes the zener and base drive currents. The zener diode dissipates minimal power because Iz is kept low at 5 mA.

Some useful formulas are:

$I_B = I_C/B = I_L/B$ (approximately)

$R1 = (V1 - V2)/(I_z + I_B)$

$P_{R1} = (V1 - V2)(I_z + I_B)$

Power Supplies

$$P_z = I_z V_z = 0.005 V_z$$

$$V_3 = V_z - V_{BE} = V_z - 0.6$$

Small variations of the input voltage causes small zener voltage fluctuations because the zener diode is not ideal. This ripple effect can be minimized by placing a capacitor in parallel with the zener diode. Since the zener current is small compared to the load current, filtering is more effective. The filtering effect of placing the capacitor in parallel with the zener diode, over the filtering effect of placing the capacitor in parallel with the load, is of the order of the transistor beta.

Computer-Aided Power Supply Component Selection

A power supply can be designed on the computer using the Basic program listing of *Table 1-1*. The three power supply circuits of the program — namely, the full-wave center-tapped circuit, the full-wave bridge circuit, and the full-wave bridge center-tapped circuit — are shown in *Figure 1-12*.

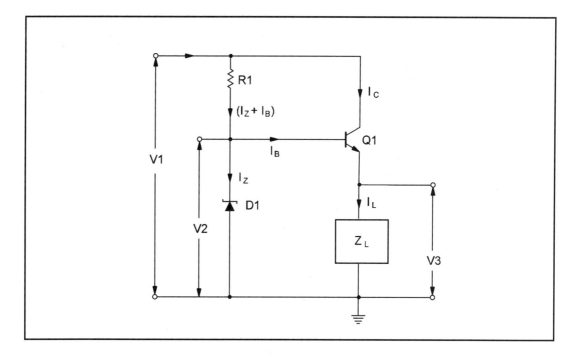

Figure 1-11. Transistor series regulator.

IC Design Projects

```
10   CLS
20   PRINT "This program designs power supplies with capacitor
     input filter"
30   PRINT "and it may be used in conjunction with program PS2
     which designs"
40   PRINT "positive and/or negative power supply regulator
     sections."
50   INPUT "Do you want screen display (S) or printer output
     (P)"; D$
60   IF D$ = "S" THEN 80
70   IF D$ = "P" THEN 500
80   CLS
85   PRINT "POWER SUPPLY TYPE"
90   PRINT "FULL WAVE BRIDGE (1)"
100  PRINT "FULL WAVE CENTER TAP (2)"
110  PRINT "FULL WAVE BRIDGE CENTER TAP (3)"
120  PRINT "ALL USING CAPACITOR INPUT FILTERING"
130  INPUT "SELECT TYPE 1,2,3 OR 0 TO QUIT"; T
135  IF T = 0 THEN END
140  ON T GOTO 1000,2000,3000
150  GOTO 80
500  CLS
510  PRINT "THE PRINTER OPTION PRINTS A TABLE OF VALUES FOR
     VARIOUS TRANSFORMERS"
520  PRINT "TYPE OF CIRCUIT:"
530  PRINT "FULL WAVE BRIDGE (1)"
540  PRINT "FULL WAVE CENTER TAP (2)"
550  PRINT "FULL WAVE BRIDGE CENTER TAP (3)"
560  PRINT "ALL USING CAPACITOR INPUT FILTERING"
570  INPUT "SELECT TYPE 1,2,3 OR 0 TO QUIT"; T
580  IF T = 0, THEN END
590  ON T GOTO 1500,2500,3500
1000 CLS
1010 PRINT "FULL WAVE BRIDGE, CAPACITOR INPUT FILTER"
1020 INPUT "SELECT TYPE 1,2,3 OR 0 TO QUIT"; T
1025 IF V = 0 THEN 80
1030 INPUT "TRANSFORMER CURRENT RATING (AMPS)"; C
1040 RV = INT(V * 1.414 - .7)
1050 DV = INT(1.414 * V + .5)
1060 CV = INT(RV)
1070 OC = INT(C/2 + .5)
1080 GOSUB 5000
1090 PRINT V; TAB(21)RV; TAB(31)DV; TAB(41)CV; TAB(52)OC
```

Table 1-1a. Basic program for power supply component selection.

```
1100 GOTO 85
1500 CLS
1510 INPUT "MINIMUM TRANSFORMER VOLTAGE"; M
1522 T$ = "FULL WAVE BRIDGE"
1524 GOSUB 6000
1530 FOR V = M TO M + 50
1540 RV = INT(V * 1.414 - .7)
1550 DV = INT(1.414 * V + .5)
1560 CV = INT(RV)
1580 LPRINT TAB(8)V; TAB(20)RV; TAB(31)DV; TAB(41)CV
1590 NEXT V
1600 GOTO 500
2000 CLS
2010 PRINT "FULL WAVE CENTER TAPPED, CAPACITOR INPUT FILTER"
2020 INPUT "TRANSFORMER RMS VOLTAGE (ENTIRE SECONDARY)"; V
2025 IF V = 0 THEN 80
2030 INPUT "TRANSFORMER CURRENT RATING (AMPS)"; C
2040 RV = INT(V * .707 - .1)
2050 DV = INT(2 * RV)
2060 CV = INT(RV)
2070 OC = INT(C/2 + .5)
2080 GOSUB 5000
2090 PRINT V; TAB(21)RV; TAB(31)DV; TAB(41)CV; TAB(52)OC
2110 GOTO85
2500 CLS
2510 INPUT "MINIMUM TRANSFORMER VOLTAGE (ENTIRE SECONDARY)"; M
2520 T$ = "FULL WAVE CENTER TAP"
2524 GOSUB 6000
2530 FOR V = M TO M + 50
2540 RV = INT(.707 * V - .1)
2550 DV = INT(2 * RV)
2560 CV = INT(RV)
2580 LPRINT TAB(8)V; TAB(20)RV; TAB(31)DV; TAB(41)CV
2590 NEXT V
2600 GOTO 500
3000 CLS
3010 PRINT "FULL WAVE BRIDGE CENTER TAP, CAPACITOR INPUT FILTER"
3020 INPUT "TRANSFORMER RMS VOLTAGE (ENTIRE SECONDARY)"; V
3025 IF V = 0 THEN 80
3030 INPUT "TRANSFORMER CURRENT RATING (AMPS)"; C
3040 RV = INT(V * .707 - .7)
3050 DV = INT(1.414 * V + .5)
3060 CV = INT(RV)
```

Table 1-1b. Basic program for power supply component selection (continued).

```
3070 OC = INT(C/1 + .5)
3080 GOSUB 5000
3090 PRINT V; TAB(18)"+/-"RV; TAB(31)DV; TAB(41)CV; TAB(49)
     "+/1"OC
3100 GOTO 85
3500 CLS
3510 INPUT "MINIMUM TRANSFORMER VOLTAGE (ENTIRE SECONDARY)"; M
3520 T$ = "FULL WAVE BRIDGE CENTER TAP"
3525 GOSUB 6000
3530 FOR V = M TO M + 50
3540 RV = INT(V * .707 - .7)
3550 DV = INT(V * 1.414 + .5)
3560 CV = INT(RV)
3580 LPRINT TAB(8)V; TAB(19)"+/-"RV; TAB(31)DV; TAB(41)CV
3590 NEXT V
3600 GOTO 500
5000 PRINT "TRANSFORMER"; TAB(20)"DC"; TAB(30)"DIODE";
     TAB(40)"CAPACITOR"; TAB(52)"DC"
5010 PRINT "VOLTAGE (RMS)"; TAB(20)"VOLTS"; TAB(30)"PRV";
     TAB(40)"VOLTAGE"; TAB(51)"CURRENT (AMPS)"
5020 RETURN
6000 LPRINT:LPRINT:LPRINT
6005 LPRINT TAB(20) T$
6010 LPRINT TAB(20)"CAPACITOR INPUT FILTER"
6015 LPRINT
6020 LPRINT TAB(5)"TRANSFORMER"; TAB(20)"DC"; TAB(30)"DIODE";
     TAB(40)"CAPACITOR"
6030 LPRINT TAB(5)"VOLTAGE (RMS)"; TAB(20)"VOLTS";
     TAB(30)"PRV"; TAB(40)"VOLTAGE"
6035 LPRINT
6040 RETURN
```

Table 1-1c. Basic program for power supply component selection (continued).

The full-wave center-tapped and full-wave bridge circuits are used for designing single-ended power supplies. The full-wave bridge center-tapped circuit is used for designing dual or bipolar power supplies.

The program operator can select either the screen as the output device when you type "s," or the printer as the output device when you type "p." If the printer is the output device, a list of component rating values is generated. Depending upon the power supply circuit selected, the appropriate list is generated as shown in *Tables 1-2, 1-3* and *1-4*.

Power Supplies

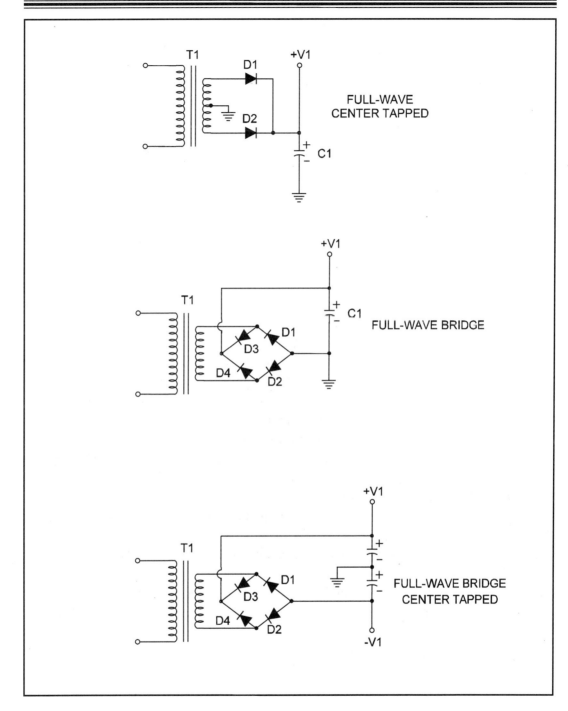

Figure 1-12. Three power supply circuits of the Basic program.

IC Design Projects

The screen output option treats the transformer as the variable element. The screen display also deals with the transformer current rating. The current rating of the transformer is derated fifty percent for reliable operation. The capacitor voltage and the diode PRV are the minimum ratings required for these components. For reliable operation, the next highest-rated component available should be used.

A "0" entry for a question returns you to the menu. A "0" entry on the menu stops the program.

Problems

Problem 1-1. Why do electronic circuits require stable power supplies?
Problem 1-2. Why are batteries not the power supply of choice?
Problem 1-3. Name the three sections of an AC-to-DC power supply.
Problem 1-4. Why does the aircraft industry use power supplies that operate at 400 hertz?
Problem 1-5. How does a transformer work?
Problem 1-6. What is the current flow through the secondary winding of a 50 VA transformer if the secondary voltage is 25 volts?
Problem 1-7. Define a step-up transformer. Define a step-down transformer.
Problem 1-8. Define an isolation transformer.
Problem 1-9. Why are regulators required in power supplies?
Problem 1-10. In essence, what is the regulator function in an AC-to-DC power supply?
Problem 1-11. What value capacitor is required for a one-volt peak-to-peak ripple voltage, when the load current is one ampere, and a full-wave rectifier circuit is used?
Problem 1-12. Design a 5-volt zener diode regulator that can deliver 50 mA of current to a load. The input voltage to the regulator is 9 volts.
Problem 1-13. Design a 5-volt transistor series regulator that can deliver one ampere of current to a load. The input voltage to the regulator is 9 volts. The transistor beta is 50.

Power Supplies

FULL WAVE CENTER TAP CAPACITOR INPUT FILTER

Transformer Voltage (RMS)	DC Volts	Diode PRV	Capacitor Voltage
2	1	2	1
3	2	4	2
4	2	4	2
5	3	6	3
6	4	8	4
7	4	8	4
8	5	10	5
9	6	12	6
10	6	12	6
11	7	14	7
12	8	16	8
13	9	18	9
14	9	18	9
15	10	20	10
16	11	22	11
17	11	22	11
18	12	24	12
19	13	26	13
20	14	28	14
21	14	28	14
22	15	30	15
23	16	32	16
24	16	32	16
25	17	34	17
26	18	36	18
27	18	36	18
28	19	38	19
29	20	40	20
30	21	42	21
31	21	42	21
32	22	44	22
33	23	46	23
34	23	46	23
35	24	48	24
36	25	50	25
37	26	52	26
38	26	52	26
39	27	54	27
40	28	56	28
41	28	56	28
42	29	58	29
43	30	60	30
44	31	62	31
45	31	62	31
46	32	64	32
47	33	66	33
48	33	66	33
49	34	68	34
50	35	70	35
51	35	70	35
52	36	72	36

Table 1-2. Basic program-generated values for full-wave center-tap power supply.

FULL WAVE BRIDGE CAPACITOR INPUT FILTER

Transformer Voltage (RMS)	DC Volts	Diode PRV	Capacitor Voltage
1	0	1	0
2	2	3	2
3	3	4	3
4	4	6	4
5	6	7	6
6	7	8	7
7	9	10	9
8	10	11	10
9	12	13	12
10	13	14	13
11	14	16	14
12	16	17	16
13	17	18	17
14	19	20	19
15	20	21	20
16	21	23	21
17	23	24	23
18	24	25	24
19	26	27	26
20	27	28	27
21	28	30	28
22	30	31	30
23	31	33	31
24	33	34	33
25	34	35	34
26	36	37	36
27	37	38	37
28	38	40	38
29	40	41	40
30	41	42	41
31	43	44	43
32	44	45	44
33	45	47	45
34	47	48	47
35	48	49	48
36	50	51	50
37	51	52	51
38	53	54	53
39	54	55	54
40	55	57	55
41	57	58	57
42	58	59	58
43	60	61	60
44	61	62	61
45	62	64	62
46	64	65	62
47	65	66	65
48	67	68	67
49	68	69	68
50	70	71	70
51	71	72	71

Table 1-3. Basic program-generated values for full-wave bridge power supply.

Power Supplies

FULL WAVE BRIDGE CENTER TAP CAPACITOR INPUT FILTER

Transformer Voltage (RMS)	DC Volts	Diode PRV	Capacitor Voltage
2	+/- 0	3	0
3	+/- 1	4	1
4	+/- 2	6	2
5	+/- 2	7	2
6	+/- 3	8	3
7	+/- 4	10	4
8	+/- 4	11	4
9	+/- 5	13	5
10	+/- 6	14	6
11	+/- 7	16	7
12	+/- 7	17	7
13	+/- 8	18	8
14	+/- 9	20	9
15	+/- 9	21	9
16	+/- 10	23	10
17	+/- 11	24	11
18	+/- 12	25	12
19	+/- 12	27	12
20	+/- 13	28	13
21	+/- 14	30	14
22	+/- 14	31	14
23	+/- 15	33	15
24	+/- 16	34	16
25	+/- 16	35	16
26	+/- 17	37	17
27	+/- 18	38	18
28	+/- 19	40	19
29	+/- 19	41	19
30	+/- 20	42	20
31	+/- 21	44	21
32	+/- 21	45	21
33	+/- 22	47	22
34	+/- 23	48	23
35	+/- 24	49	24
36	+/- 24	51	24
37	+/- 25	52	25
38	+/- 26	54	26
39	+/- 26	55	26
40	+/- 27	57	27
41	+/- 28	58	28
42	+/- 28	59	28
43	+/- 29	61	29
44	+/- 30	62	30
45	+/- 31	64	31
46	+/- 31	65	31
47	+/- 32	66	32
48	+/- 33	68	33
49	+/- 33	69	33
50	+/- 34	71	34
51	+/- 35	72	35
52	+/- 36	74	36

Table 1-4. Basic program-generated values for full-wave bridge center-tap power supply.

Chapter 2
◆ IC Regulators ◆

Most electronic circuits require a stable DC voltage power supply. The regulator reduces the AC component of an AC-to-DC power supply output voltage. There are two types of integrated circuit regulators — the switching regulator and the linear regulator.

The switch-mode power supply is small in size, yet it has a high conversion efficiency. The switch-mode power supply is popular for commercial use, such as in computers. The linear power supply, on the other hand, is the choice for hobbyists because it is easy to design and build.

The IC regulator eliminates the need to design from scratch, using individual operational amplifiers, precision voltage references and hand-selected transistors.

Linear IC Regulator

Most linear IC regulators are DC operational amplifiers. The DC operational amplifier is a basic analog building block, and it makes good use of the well-matched characteristics of monolithic components. It is difficult and expensive to duplicate monolithic characteristics with discrete components.

A basic voltage regulator circuit is shown in *Figure 2-1*. An operational amplifier is used to compare a reference voltage with a fraction of the output voltage. The operational amplifier also controls a series-pass element to regulate the output, depending on the differential voltage between its input terminals.

The current handling ability of monolithic circuits is limited because of the large physical die size of high-current transistors. Power dissipation is also a problem because there are no readily-available multilead power packages for integrated circuits. Higher output current capability and improved load regulation can be obtained by using external transistors. The output current is now limited by the external transistor power dissipation and current handling capabilities. Using external transistors as series-pass ele-

IC Design Projects

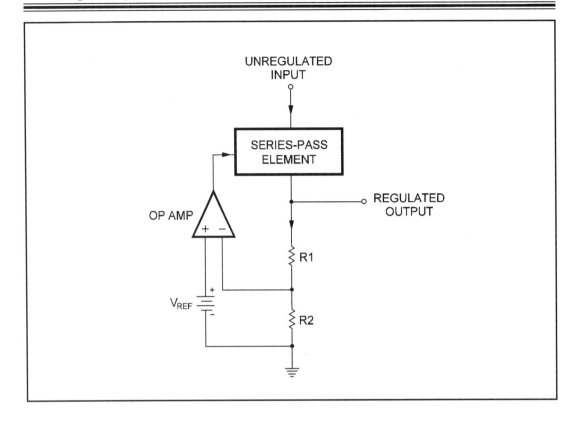

Figure 2-1. Basic linear regulator circuit.

ments reduces the internal dissipation of the integrated circuit, thus preventing the temperature drift caused by internal heating.

A well-regulated power supply is required for most electronic circuits. The suitability of the design of the regulators to monolithic construction can be shown by the fact that a regulator can be built on a 38-mil. square silicon die: a size that is comparable to a silicon transistor. The small size helps to achieve high yields, which are necessary to obtain low manufacturing costs and to ensure off-the-shelf availability.

Switch-Mode Integrated Circuit Regulator

The series-pass element of a linear regulator operates as a variable resistance, which drops an unregulated input voltage down to a fixed output voltage. The element is usually a transistor that must be able to drop the voltage difference. When the load

IC Regulators

current is high, the power dissipated by the series pass element — usually a transistor — can become excessive.

Switch-mode or switching regulators are capable of high-efficiency operation even with large differences between the input and output voltages. The switching regulator acts as a continuously-variable power converter.

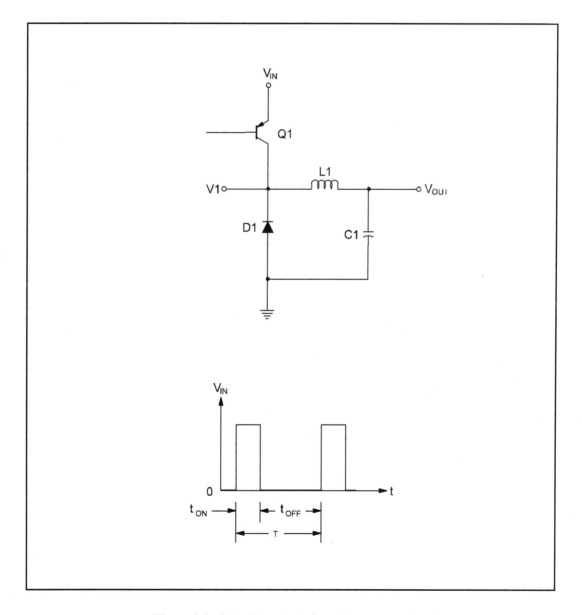

Figure 2-2. Switching circuit for voltage conversion.

Switching regulators are therefore useful in battery-powered equipment in which the required output voltage is much lower than the battery voltage. Switching regulators are also useful in space vehicles because the conservation of power is extremely important. They are economical choices for commercial and industrial applications because the increased efficiency reduces the cost of the series-pass transistors and simplifies the required heat sinking.

The switching regulator is more complex than the linear regulator, and has a higher output ripple factor. Linear regulators respond more quickly to load transients than switching regulators. Both types of regulators reject line transients equally well. Switching regulators throw current transients back to the unregulated input supply. The throwback transient currents can be larger than the maximum load current; therefore, adequate filtering is required.

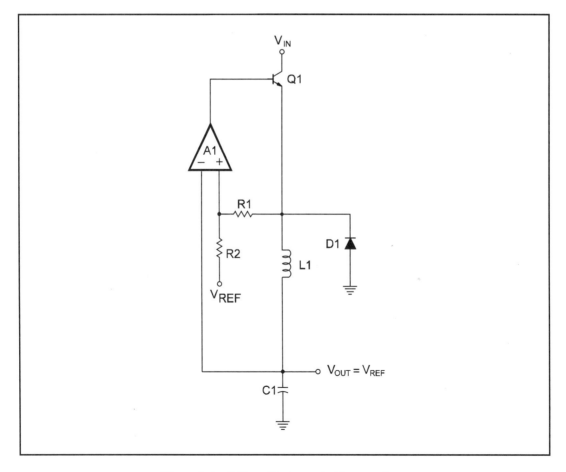

Figure 2-3. Self-oscillating switching regulator.

IC Regulators

A switching circuit for voltage conversion is shown in *Figure 2-2*. Transistor Q1 is turned on and off by a pulse waveform. Diode D1 provides a path for the inductor current when Q1 turns off. The collector voltage waveform of Q1 and V1 is shown in *Figure 2-2*. The output of the LC filter is the average value of the switched waveform, Vin. Neglecting the voltage drops across the transistor and the diode, the output voltage is Vout = Vin(t_{ON}/T).

The output voltage is independent of the load current. Changes in the input voltage of a switching regulator are compensated for by varying the duty cycle of the switched waveform.

A self-oscillating switching regulator circuit is shown in *Figure 2-3*. It produces the duty cycle control. A reference voltage, Vref (equal to the required output voltage), is applied to one input of an operational amplifier, A1. The operational amplifier drives the switch transistor. The resistive divider, which is arranged so that R1 >> R2, provides a small amount of positive feedback at high frequencies, causing the circuit to oscillate. At lower frequencies, where the attenuation of the LC filter is less than the attenuation of the resistive divider, there is negative feedback through the inverting input of the operational amplifier.

When the circuit is first turned on, the output voltage is less than the reference voltage. The switch transistor turns on and current flows through R1. The voltage on the noninverting input of the operational amplifier is slightly higher than the reference voltage. The output voltage rises to the voltage on the noninverting input of the operational amplifier. The operational amplifier goes into its active region and the switch turns off. The reference voltage is lowered by feedback through resistor R1. The circuit remains off until the output voltage drops to the lower voltage. The output oscillates around the reference voltage. The amplitude of oscillation (or the output ripple) is nearly equal to the voltage fed back through R1 to R2. The output ripple can be made quite small.

The switching regulator is not protected from overloads or short-circuited output. Providing short-circuit protection is difficult because it is necessary to keep the regulator switching when the output is short-circuited. Power dissipation can become excessive even though the current flow is limited. The output current is limited by the discrete components, not by the basic design or by the IC.

IC Design Projects

TYPE NUMBER	VOLTAGE
MC7805 MC7806 MC7808	5 volts 6 volts 8 volts
MC7812 MC7815 MC7818	12 volts 15 volts 18 volts
MC7824 MC7902 MC7905	24 volts -2 volts -5 volts
MC7905.2 MC7906 MC7908	-5.2 volts -6 volts -8 volts
MC7912 MC7915 MC7918	-12 volts -15 volts -18 volts
MC7924 LM340K-5.0 LM340K-6.0	-24 volts 5 volts 6 volts
LM340K-8.0 LM340K-12 LM340K-15	8 volts 12 volts 15 volts
LM340K-18 LM340K-24	18 volts 24 volts

Table 2-1. MC7800, LM340 and MC7900 series-type number and voltage.

Several integrated voltage regulators are available to the power supply designer. There are positive voltage regulators and negative voltage regulators. There are fixed voltage regulators and adjustable voltage regulators.

MC7800 and MC7900 Regulators

The MC7800 is a three-terminal, positive, fixed voltage-integrated circuit regulator. These regulators employ internal current limiting, thermal shutdown and safe-area compensation. The MC7900 is a three-terminal, negative, fixed voltage-integrated circuit regulator. Both series are available with various output voltage options, as listed in *Table 2-1*.

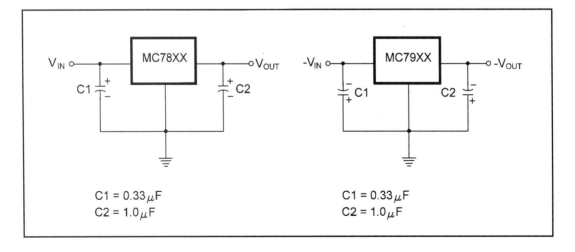

Figure 2-4. MC7800 and MC7900 standard application circuit.

IC Regulators

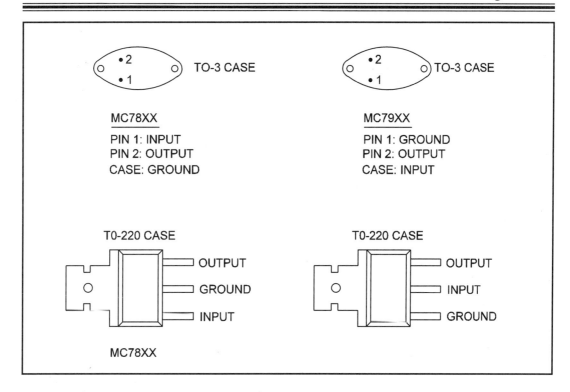

Figure 2-5. MC7800 and MC7900 series packages.

The MC7800 and MC7900 series require no external components. A common ground is required between the input and output voltages, as shown in *Figure 2-4*. The input voltage must be at least two volts higher than the output voltage. Capacitor C1 is required if the regulator is located far from the power supply. Capacitor C2 is not required for stability; however, it improves the transient response. These devices can be used with external components to obtain adjustable voltages and currents.

Both series are available in a TO-220 plastic package and in a TO-3 metal package, as shown in *Figure 2-5*. The TO-220 package can dissipate about 100 milliamperes while the TO-3 package can dissipate about one ampere, with adequate heat sinking.

The unregulated input voltage is fed to the control element and the reference voltage circuit, as shown in *Figure 2-6*. The output voltage is sampled and fed into one of the error amplifier inputs. The other error amplifier input is connected to the reference voltage. When the error amplifier senses a difference between the reference and sampling voltages, it acts upon the control element to correct the error by dropping a greater

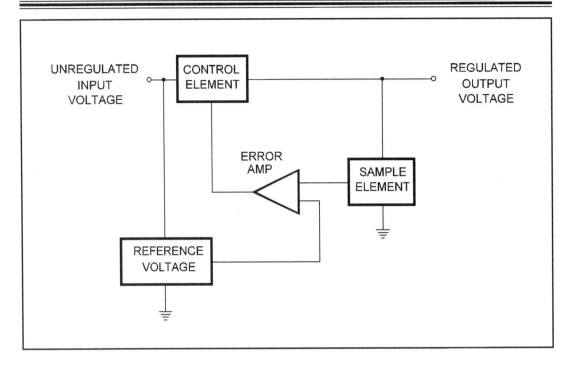

Figure 2-6. Block diagram of MC7800 and MC7900 voltage regulator.

portion of the input voltage across the control element. The control element is a transistor which acts as a variable resistor, in which the resistance is controlled by the error amplifier.

The unregulated input voltage must exceed the regulator's output voltage by at least two volts. The input voltage, however, must not be large enough to exceed the regulator power dissipation specification. Both series have an internal protection circuit, a current limiting circuit, and a safe-area protection circuit. The safe-area protection circuit limits the regulator output voltage when the input voltage is too high, ensuring that the pass transistor control element operates within its allowed voltage and current ranges.

Some design applications for the MC7800 and MC7900 series are shown in *Figure 2-7*. The LM340 series three-terminal positive voltage regulator is virtually identical to the MC7800 series three-terminal positive voltage regulator. The LM340 series has line and load regulation factors that are better than those of the MC7800 series by a factor of two. The output voltage options of the LM340 series regulator are listed in *Table 2-1*.

IC Regulators

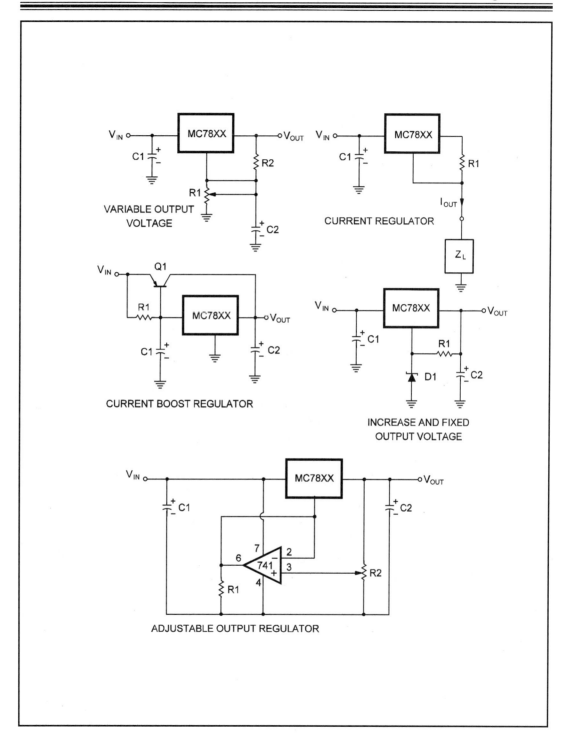

Figure 2-7. Some design applications for MC7800 and MC7900 series.

LM723 IC Voltage Regulator

The LM723 is a positive or negative voltage regulator which can deliver 150 milliamperes into a load. The output current capability can be increased to several amperes by using one or more external series-pass transistors. The LM723 has a good ripple rejection factor, excellent load and line regulation factors, and a built-in current limiting facility.

The LM723 consists of a voltage reference amplifier, a temperature-compensated reference voltage zener diode, a second zener diode, an error amplifier, a current limiter transistor, and a series-pass transistor, as shown in *Figure 2-8*. The LM723 regulator pin assignment is also shown in *Figure 2-8*. There are no connections required to pins 1, 8 and 14 of the LM723 voltage regulator.

One of the internal *npn* transistors (current limiter transistor) has its collector connected to the base of the series-pass transistor. The base (current limit or CL) is externally available at pin 2, and the emitter (current sense or CS) is externally available at pin 3. Current limiting is achieved by connecting a resistor between pins 2 and 3. When the load current is high enough to cause a voltage drop of 0.65 volts across the resistor, the current limiter transistor turns on and the drive to the series-pass transistor is either reduced or cut off. This reduces the output current flow down to the required level: $R_{SC} = V_{BE}/I_{OUT}(MAX) = 0.65/I_{OUT}(MAX)$.

The inverting and noninverting inputs to the error amplifier are available at pins 4 and 5, respectively. A reference voltage can be applied to the noninverting input of the error amplifier. A voltage derived from the output voltage of the power supply is applied to the inverting input of the error amplifier. The error amplifier detects the voltage difference between the reference and output voltages, and controls the internal *npn* series-pass transistor.

The reference voltage is derived from a constant current source and is available at pin 6. This reference voltage is connected to the noninverting input (pin 5) of the error amplifier.

If pin 7 (V-) of the regulator is connected to ground, then the minimum output voltage of an adjustable power supply is about two volts. If a minimum voltage of zero volts is required, then pin 7 should be connected to a -3.6 volt source.

IC Regulators

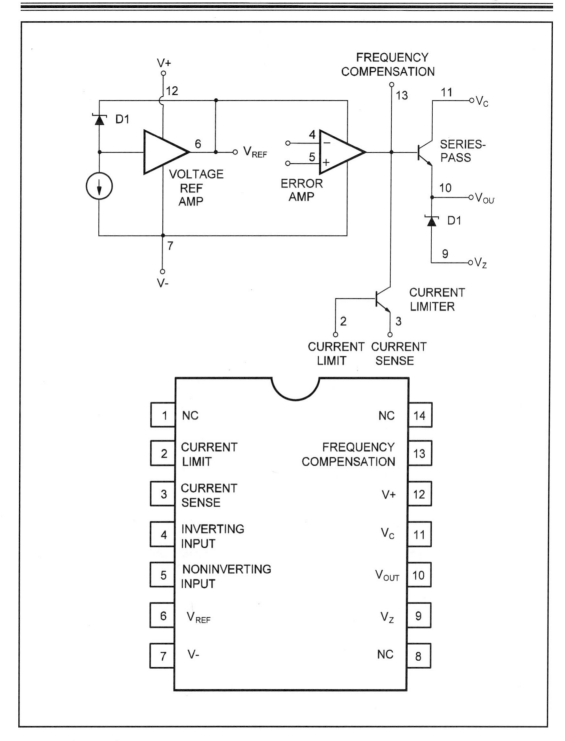

Figure 2-8. LM723 internal circuits and pin assignment.

IC Design Projects

There is a second zener diode connected to the emitter of the series-pass transistor. The anode of the second zener diode is about the same as that of the temperature-compensated reference voltage zener diode. The second zener diode is usually used to offset the output voltage when the LM723 is configured as a negative voltage regulator.

The emitter and the collector of the internal series-pass transistor are available at pins 10 and 11; Vout and V+, respectively. In most circuits, pins 11 and 12 (Vc and V+) are connected to a power supply.

The voltage difference between pins 7 and 12 (V- and V+) cannot exceed 40 volts. The voltage applied to pin 12 must also be at least 2.5 volts greater than the required regulated output voltage.

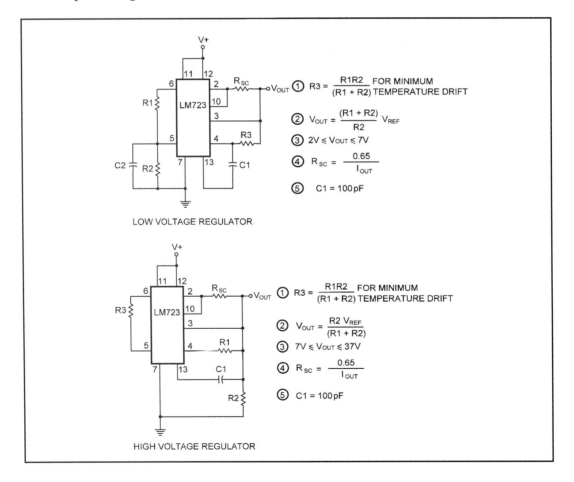

Figure 2-9. LM723 basic low-voltage and high-voltage regulators.

IC Regulators

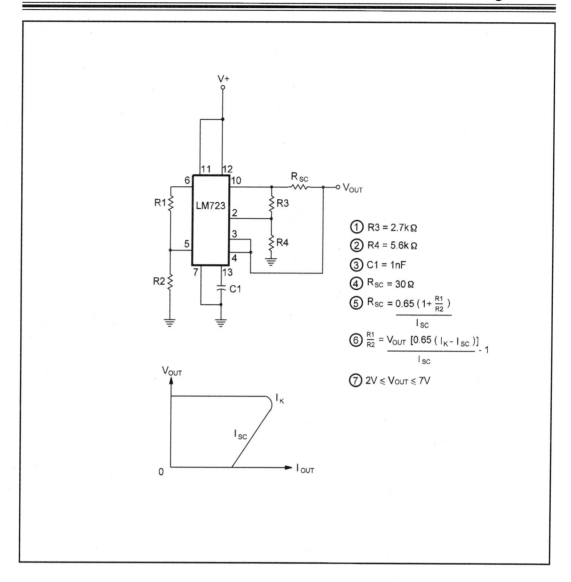

Figure 2-10. Foldback current limiting circuit.

The connection between the error amplifier output and the base of the series-pass transistor is available at pin 13 (frequency compensation). The error amplifier of the LM723 is not internally frequency compensated. An external capacitor connected between pin 13 and the inverting input (pin 4) stabilizes the error amplifier. The capacitor may be connected between pin 13 and the V- input (pin 7) to stabilize the error amplifier.

IC Design Projects

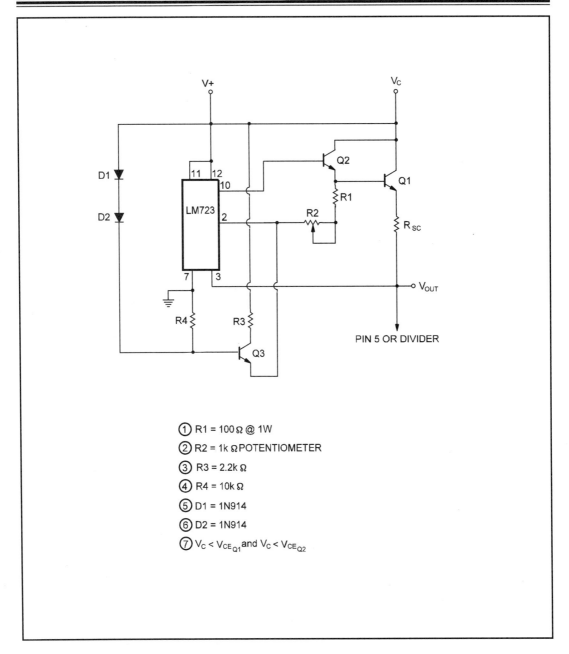

Figure 2-11. Variable-current limiting circuit.

Two basic positive voltage regulator circuits are shown in *Figure 2-9*. The low voltage regulator can be designed for an output of 2 to 7 volts, while the high voltage regulator can be designed for an output of 7 to 37 volts.

IC Regulators

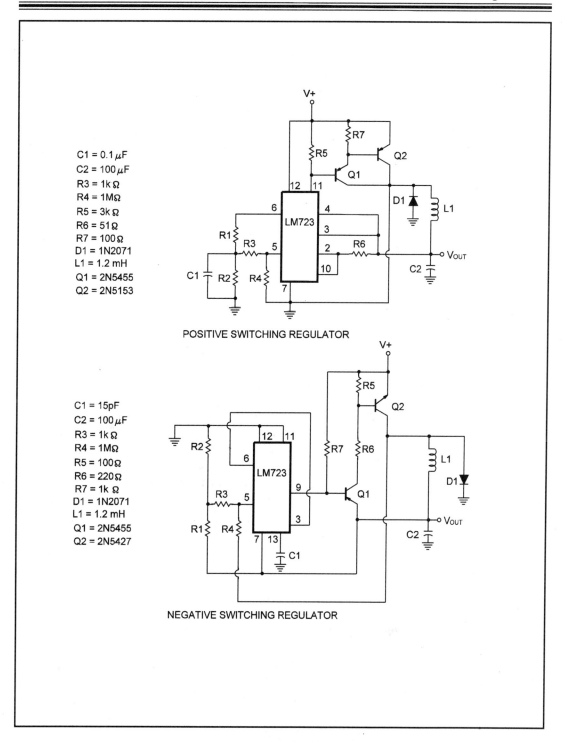

Figure 2-12. *Switching regulator circuits.*

IC Design Projects

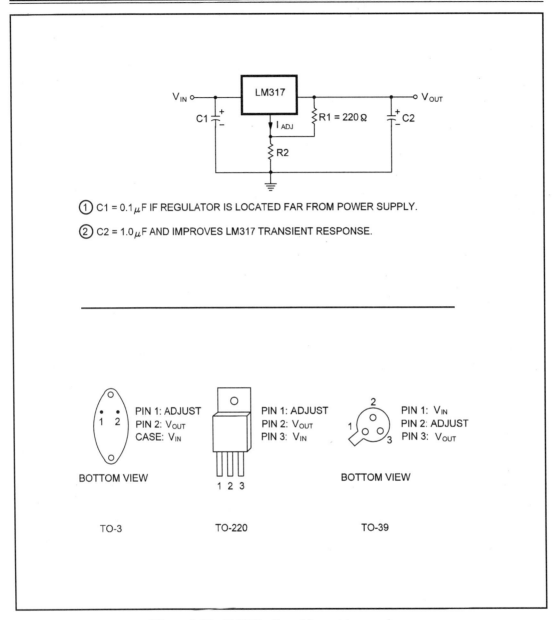

Figure 2-13. LM317 adjustable positive regulator.

The LM723 should never be operated without the external current limiting resistor. A foldback current-limiting circuit is shown in *Figure 2-10*. This circuit is suited to fixed-voltage supplies and limited-adjustability supplies. The current limiting point changes with the output voltage, with foldback limiting as shown in *Figure 2-10*. A variable-current limiting circuit is shown in *Figure 2-11*. A positive switching regulator circuit and a negative switching regulator circuit are shown in *Figure 2-12*.

IC Regulators

CALCULATIONS FOR 317 SERIES REGULATORS

Output Volts	Calculated R2	5% R2	Output V 5% R2	Min Input V	Max Input V
2	132	130	1.9	4.5	14
3	308	300	2.9	5.5	15
4	484	470	3.9	6.5	16
5	660	680	5.1	7.5	17
6	836	820	5.9	8.5	18
7	1012	1000	6.9	9.5	19
8	1188	1200	8	10.5	20
9	1364	1300	8.600001	11.5	21
10	1540	1500	9.7	12.5	22
11	1716	1800	11.4	13.5	23
12	1892	1800	11.4	14.5	24
13	2068	2000	12.6	15.5	25
14	2244	2200	13.7	16.5	26
15	2420	2400	14.8	17.5	27
16	2596	2700	16.5	18.5	28
17	2772	2700	16.5	19.5	29
18	2948	3000	18.2	20.5	30
19	3124	3000	18.2	21.5	31
20	3300	3300	20	22.5	32
21	3476	3600	21.7	23.5	33
22	3652	3600	21.7	24.5	34
23	3828	3900	23.4	25.5	35
24	4004	3900	23.4	26.5	36
25	4180	4300	25.6	27.5	37
26	4356	4300	25.6	28.5	38
27	4532	4700	27.9	29.5	39
28	4708	4700	27.9	30.5	40
29	4884	4700	27.9	31.5	40
30	5060	5100	30.2	32.5	40
31	5236	5100	30.2	33.5	40
32	5412	5600	33	34.5	40
33	5588	5600	33	35.5	40
34	5764	5600	33	36.5	40
35	5940	6200	36.4	37.5	40
36	6116	6200	36.4	38.5	40
37	6292	6200	36.4	39.5	40

Table 2-2. Resistors required for LM317 regulator designs.

LM317 and LM337 Adjustable Regulators

The MC7800 and MC7900 series regulators are only available in limited voltages. If other voltages are required, then the LM317 and LM337 integrated circuit regulators must be used. The LM317 is a three-terminal positive voltage regulator, and the LM337 is a three-terminal negative voltage regulator.

IC Design Projects

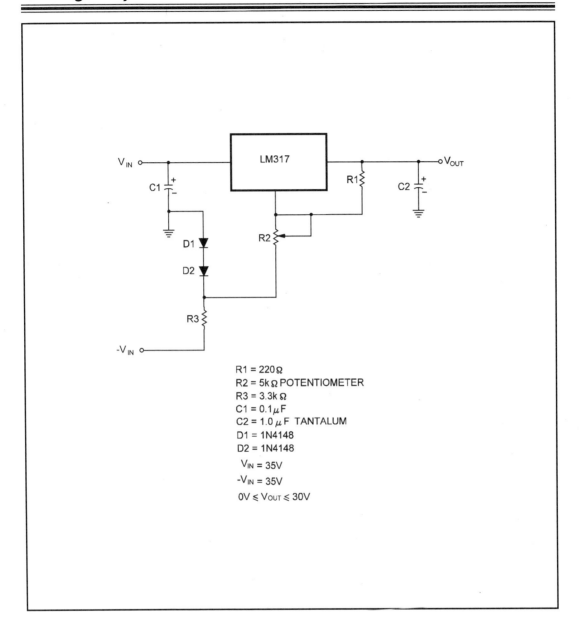

Figure 2-14. A full-variable positive voltage regulator.

The LM317 maintains a nominal 1.25 volt reference (Vref) between its output and adjustable terminals. The reference voltage is converted to a programming current (Iprog) by R1, as shown in *Figure 2-13*. The constant programming current flows through R2 to ground. The output voltage is Vout = Vref(1 + [R2/R1]) + I_{ADJ}R2

IC Regulators

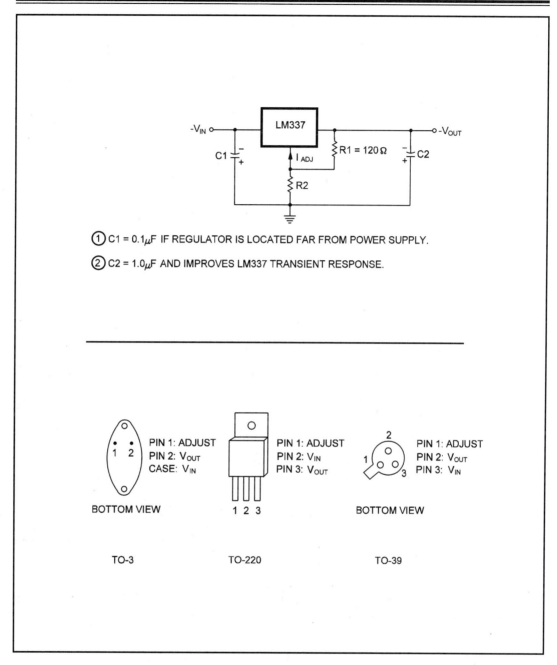

Figure 2-15. LM337 adjustable negative regulator.

R1 = 220 ohms. The TO-3 and TO-220 packages can deliver 1.5 amperes to the load if the regulator is properly heat sunk. The TO-39 package can deliver 0.5 amperes to a load.

CALCULATIONS FOR 337 SERIES REGULATORS

Output Volts	Calculated R2	5% R2	Output V 5% R2	Min Input V	Max Input V
-2	72	75	-2	-4.5	-14
-3	168	160	-2.9	-5.5	-15
-4	264	270	-4	-6.5	-16
-5	360	360	-5	-7.5	-17
-6	456	470	-6.1	-8.5	-18
-7	552	560	-7	-9.5	-19
-8	648	620	-7.7	-10.5	-20
-9	744	750	-9	-11.5	-21
-10	840	820	-9.7	-12.5	-22
-11	936	910	-10.7	-13.5	-23
-12	1032	1000	-11.6	-14.5	-24
-13	1128	1100	-12.7	-15.5	-25
-14	1224	1200	-13.7	-16.5	-26
-15	1320	1300	-14.7	-17.5	-27
-16	1416	1500	-16.8	-18.5	-28
-17	1512	1500	-16.8	-19.5	-29
-18	1608	1600	-17.9	-20.5	-30
-19	1704	1800	-20	-21.5	-31
-20	1800	1800	-20	-22.5	-32
-21	1896	1800	-20	-23.5	-33
-22	1992	2000	-22	-24.5	-34
-23	2088	2000	-22	-25.5	-35
-24	2184	2200	-24.1	-26.5	-36
-25	2280	2200	-24.1	-27.5	-37
-26	2376	2400	-26.2	-28.5	-38
-27	2472	2400	-26.2	-29.5	-39
-28	2568	2700	-29.3	-30.5	-40
-29	2664	2700	-29.3	-31.5	-40
-30	2760	2700	-29.3	-32.5	-40
-31	2856	3000	-32.5	-33.5	-40
-32	2952	3000	-32.5	-34.5	-40
-33	3048	3000	-32.5	-35.5	-40
-34	3144	3000	-32.5	-36.6	-40
-35	3240	3300	-35.6	-37.5	-40
-36	3336	3300	-35.6	-38.5	-40
-37	3432	3300	-35.6	-39.5	-40

Table 2-3. Resistors required for LM337 regulator designs.

The adjustment current is less than 100 microamperes; therefore, with a very small error, the output voltage is Vout = Vref(1 + [R2/R1]), with R1 = 220 ohms.

You may have to calculate R2 for a specific output voltage. With R1 = 220 ohms, R2 = R1([Vout/Vref] - 1).

The quiescent operating current flows through the output terminal. A minimum load current is therefore required. The output voltage increases if there is no load across its output terminals.

IC Regulators

The LM317 is a floating regulator. Only the voltage difference across its input and output terminals is important to its performance. Operation at high voltages with respect to ground is possible.

The LM317 has built-in foldback current limiting and thermal protection circuits. Its input voltage range is 4 to 40 volts, and its output voltage range is 1.5 to 37.5 volts. *Table 2-2* lists the resistors required for LM317 adjustable positive regulator designs.

A fully-variable positive voltage regulator is shown in *Figure 2-14*. Resistor R3 limits the current flow through diodes D1 and D2 to 10 milliamperes. One "leg" of potentiometer R2 is held at a constant -1.25 volts. The minimum output voltage of the variable positive voltage regulator is zero volts.

The LM337 maintains a nominal -1.25 volt reference (Vref) between its output and adjustment terminals. The reference voltage is converted to a programming current (Iprog) by R1, as shown in *Figure 2-15*. The constant programming current flows through R2 to ground. The output voltage is -Vout = -Vref(1 + [R2/R1]) + I_{ADJ}R2.

R1 = 120 ohms. The TO-3 and TO-220 packages can deliver 1.5 amperes to the load if the regulator is properly heat sunk. The TO-39 package can deliver 0.5 amperes to a load.

The adjustment current is less than 100 microamperes; therefore, with very little error, the output voltage is -Vout = -Vref(1 + [R2/R1]). R1 = 120 ohms.

The designer may be required to calculate R2 for a specific output voltage. R1 = 120 ohms. R2 = R1([Vout/Vref] - 1).

The quiescent operating current flows through the output terminal. A minimum load current is therefore required. The output voltage increases if there is no load across its output terminals.

The LM337 is a floating regulator. Only the voltage difference across its input and output terminals is important to its performance. Operation at high voltages with respect to ground is possible.

The LM337 has built-in foldback current limiting and thermal protection circuits. Its input voltage range is -4 to -40 volts, and its output voltage range is -1.5 to -37.5 volts. *Table 2-3* lists the resistors required for LM337 adjustable positive regulator designs.

IC Design Projects

Problems

Problem 2-1. What is the function of a voltage regulator?

Problem 2-2. Name two types of regulators.

Problem 2-3. What are some features of a switch-mode power supply?

Problem 2-4. What are some features of a linear power supply?

Problem 2-5. What is the main building block of an integrated circuit linear regulator?

Problem 2-6. How does a linear regulator work?

Problem 2-7. How is it possible to design power supplies that can deliver higher currents than that specified by the manufacturer for the integrated circuit regulator?

Problem 2-8. In what applications are switching regulators useful?

Problem 2-9. Calculate the output voltage of a switching regulator if the input voltage is 20 volts, t_{ON} = 10 usec and T = 2- usec.

Problem 2-10. What are the internal building blocks of an LM723 integrated circuit regulator?

Problem 2-11. A voltage regulator using an LM723 is required not to pass currents in excess of one ampere. Calculate R_{cs}.

Problem 2-12. If pin 7 of an LM723 regulator is connected to ground, what is the minimum output voltage of the regulator?

Problem 2-13. To what voltage should pin 7 of an LM723 regulator be connected for the minimum output voltage of the regulator to be zero volts?

Problem 2-14. What is the purpose of the second zener diode of an LM723 regulator?

Problem 2-15. An LM317 regulator has a reference voltage of 1.25 volts. If R1 = 220 ohms and R2 = 1000 ohms, calculator Vout.

Problem 2-16. An LM337 regulator has a reference voltage of -1.25 volts. If R1 = 120 ohms and Vout = -12 volts, calculate R2.

Chapter 3

♦ Battery Charger ♦

Nickel-cadmium rechargeable batteries come in a wide variety of terminal voltages and current ratings. This battery charger can charge most nickel-cadmium rechargeable batteries. The charging rate is switch-selectable from 50 mA/h to 2500 mA/h. Any battery with a terminal voltage of up to 20 volts is automatically accommodated. There is no voltage selection required.

The charging time required for a fully discharged battery is approximately fourteen hours. Overcharging at the correct charging rate will not damage a nickel-cadmium rechargeable battery. The battery should never be charged at a higher rate than its specified maximum charging rate.

Circuit Operation

The battery charger consists of a step-down transformer, a full-wave rectifier and a current regulator, as shown in the schematic in *Figure 3-1*. The current regulator is switchable.

Transformer T1 steps down the household AC line voltage to a more usable, lower AC voltage. The transformer secondary voltage required is dependent upon the battery or batteries to be charged, as listed in *Table 3-1*. The current rating of the transformer should be greater than the maximum battery charging rate by a factor of ten. A transformer with a single winding secondary of half voltage may be used with a bridge rectifier, as listed in *Table 3-1*.

Diodes D1 and D2 rectify the low AC voltage into a pulsating DC voltage. Capacitor C1 reduces the AC component of the pulsating DC voltage. The pulsating DC voltage is now a fluctuating DC voltage. *Table 3-1* shows the voltage rating of the capacitor.

Transistor Q1 is configured as a current regulator. In a current regulator, the current is constant regardless of the changes in the load impedance. The output voltage changes to maintain a constant load current. The load current is selectable by rotary switch S2.

IC Design Projects

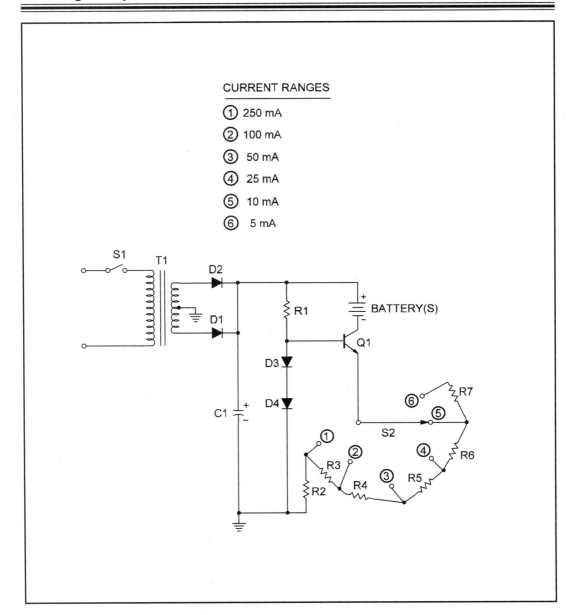

Figure 3-1. Schematic of the battery charger.

The transistor is biased by diodes D3 and D4. The base of Q1 is maintained at 1.2 volts. There is a voltage drop of 0.6 volts across the selected resistor network, R2 - R7, because there is also a voltage drop of 0.6 volts across the base-emitter junction of the transistor. The emitter current of Q1 is therefore $I_E = (V_B - V_{BE})/R_X = 0.6/R_X$, where R_X is the selected portion of the resistor network consisting of R2 - R7. The collector

Battery Charger

Battery(s) Terminal Volt. (V)	Transformer Secondary Volt. for Full-Wave Rectifier	Transformer Secondary Volt. for Full-Wave Bridge Rectifier	Minimum Capacitor Volt. Rating	R1
1.25 - 3.75	12.6 V.C.T.	6.3V	9.0V	1.8kΩ
5.0 - 10.0	25.2 V.C.T.	12.6V	18.0V	2.2kΩ
11.25 - 20.0	40.0 V.C.T.	20.0V	28.0V	3.9kΩ
21.0 - 30.0	60.0 V.C.T.	30.0V	42.0V	5.6kΩ

Table 3-1. Component specifications.

current that charges the battery is approximately equal to the emitter current. The collector current remains constant as long as there is at least a one-volt drop between the collector and emitter terminals of transistor Q1. Resistor R1 limits the current flow through the transistor. The value of R1 is listed in *Table 3-1*.

The constant current source of this battery charger can be used to charge any battery up to 20 volts. If the 20-volt capability is not required, a different transformer may be used, as detailed in *Table 3-1*.

If only a single charging rate is required, a single resistor may be used to replace the R2 - R7 resistor network. The resistor value in ohms should be R = 6000/I, where I is the maximum charging rate of the battery in mA/h.

Construction

The battery charger is very easy to build. It may be built on a piece of perfboard or on a pair of tag strips. Range resistors R2 - R7 may be mounted directly on the range switch, S2. The parts list for the battery charger is listed in *Table 3-2*.

R1:	See Table 3-1
R2:	2.2 ohms
R3:	3.9 ohms
R4:	5.6 ohms
R5:	12 ohms
R6:	39 ohms
R7:	56 ohms
C1:	1000 uF; for minimum working voltage, see Table 3-1
D1-D4:	1N4004
Q1:	2N3055
T1:	For secondary voltage, see Table 3-1. Secondary current: 1/2A
S1:	SPST
S2:	SP6T

Table 3-2. Parts list for the battery charger.

The transistor dissipates some heat on the higher charging rates. The transistor should therefore be mounted on a heat sink. A piece of aluminum can be bent into shape to serve as a heat sink. Either the transistor should be electrically insulated from the heat sink using a mica washer and insulating bushes, or the aluminum heat sink should be electrically insulated from the battery charger cabinet.

Testing and Use

Do not install a rechargeable battery yet. The minimum capacitor voltage should be measured across the capacitor. If there is no voltage, verify the polarity of diodes D1 and D2. If the capacitor voltage is correct but reversed in polarity, then diodes D1 and D2 are installed backward. When the capacitor voltage is correct, install the rechargeable battery.

The base voltage of transistor Q1 should be 1.2 volts. If it is very high, then diodes D3 and D4 are installed backward. The voltage drop across the selected portion of the resistor network, R2 - R7, should be 0.6 volts. Most construction errors are caused by either solder shorts or by cold solder joints.

Chapter 4
◆ Bipolar Power Supply ◆

Most linear ICs require connections to both a positive and a negative power supply. These linear ICs require a dual or bipolar power supply because they have one or more differential amplifiers in their circuits. Linear integrated circuits usually require symmetrical bipolar power supplies. Both bipolar power supply voltages are referenced to a common or a ground voltage.

The MC7800 and MC7900 series fixed-voltage regulators are the heart of this generic dual power supply project. The bipolar power supply is ideal for use as a power supply for most operational amplifier circuits.

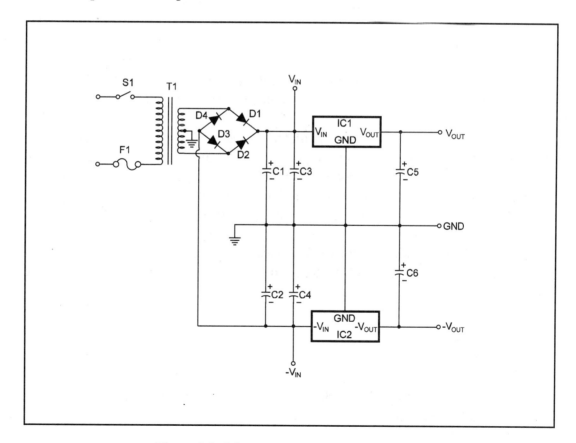

Figure 4-1. Schematic of the bipolar power supply.

(C5-C6 Working Voltage) Required Vout	Transformer Secondary Voltage	Input Voltage (C1-C4 Working Voltage)	IC1	IC2
± 5V	12.6 V.C.T.	± 9V	7805	7905
± 6V	12.6 V.C.T.	± 9V	7806	7906
± 8V	18.0 V.C.T.	± 13V	7808	7908
± 12V	25.2 V.C.T.	± 18V	7812	7912
± 15V	25.2 V.C.T.	± 18V	7815	7915
± 18V	36.0 V.C.T.	± 25V	7818	7918
± 24V	40.0 V.C.T.	± 28V	7824	7924

Table 4-1. Component specifications.

Circuit Description

The schematic of the generic bipolar power supply is shown in *Figure 4-1*. Transformer T1 steps down the AC household line voltage to a more usable, lower AC voltage. Rectifier diodes D1 - D4 convert the lower AC voltage to a pulsating DC voltage. Filter capacitors C1 and C2 convert the pulsating DC voltage to a fluctuating DC voltage by reducing the AC component of the DC voltage. Capacitors C3 and C4 are decoupling capacitors. They allow unwanted signals to be bypassed to ground.

The positive fixed-voltage output is regulated by the MC78xx positive fixed-voltage regulator. The negative fixed-voltage output is regulated by the MC79xx negative fixed-voltage regulator. Both voltage regulators require the unregulated input voltage to be at least two volts higher than the required regulated output voltage. Capacitors C5 and C6 improve the transient response of the voltage regulators.

The output voltages required may vary from circuit to circuit. *Table 4-1* lists the transformers required for the various MC78xx and MC79xx positive and negative voltage regulators.

Construction

The bipolar power supply is very easy to build. It may be built on a piece of perfboard or on a printed circuit board. The diodes and capacitors must be properly oriented. The parts list is given in *Table 4-2*.

Bipolar Power Supply

Seven possible symmetrical bipolar power supplies are listed in *Table 4-1*. The transformer secondary voltage specification, the appropriate regulators, and the capacitor working voltages are also listed in *Table 4-1*.

The MC78xx and MC79xx regulators should be properly heat sunk. A piece of aluminum bent into shape can serve as a heat sink. Either the regulators should be electrically insulated from the heat sinks, or the heat sinks should be electrically insulated from the power supply cabinet.

Testing and Use

The bipolar power supply should work the first time it is powered up. If not, verify that the capacitors and diodes are properly oriented. Check for cold solder joints and solder bridges. Verify that the positive and negative voltage regulators are properly installed. The bipolar power supply should provide years of reliable service.

```
D -D4:   1N5404
T1:      See Table 4-1; secondary current is 1A
C1-C2:   2200 uF; see Table 4-1 for working voltage
C3-C4:   0.33 uF tantalum; see Table 4-1 for working voltage
C5-C6:   1.0 uF tantalum; see Table 4-1 for working voltage
IC1:     See Table 4-1
IC2:     See Table 4-1
S1:      SPST
F1:      1/2A slow blow fuse
```

Table 4-2. Parts list for the bipolar power supply.

Chapter 5
♦ 5-Volt Power Supply ♦

Transistor-transistor logic (TTL) circuits require a very well-regulated 5-volt power supply. This power supply project incorporates many features that are found only on expensive power supplies.

The output voltage is very well-regulated regardless of large fluctuations in household line voltage. This power supply can deliver 3.5 amperes of current into a load. Over-voltage protection is provided. It will short out the output if the output voltage exceeds 6.8 volts.

Foldback current limiting is provided. It limits the output current to a safe level whenever an overload or a short circuit is connected to the output terminals of the power supply.

Remote sense terminals are available and can compensate for voltage drops in the leads connecting the power supply to the load. A 5-ampere meter monitors the output current. The brains of the power supply is an LM723 IC voltage regulator. The specifications for the 5-volt supply are listed in *Table 5-1*.

Circuit Description

The schematic of the 5-volt power supply is shown in *Figure 5-1*. Transformer T1 steps the household line voltage to a more usable, lower AC voltage. Diodes D1 - D4

Input Voltage:	105-135 VAC
Output Voltage:	5.0 VDC (+/- 0.5V adjustable)
Voltage Regulation:	Less than 0.05% from 105-135 VAC
Voltage Regulation:	Less than 0.05% at full load
Ripple:	Less than 1 mV at full load
Overvoltage Protection:	Activates at 6.8 VDC (+/- 0.68V)
Current Limit:	3.5 Amps (+/- 0.525A)
Short Circuit Current:	0.5 Amps (adjustable), 1.5A (max)
Sense Lines:	0.5V per load line to 2.5 Amps
	0.2V per load line to 3.5 Amps

Table 5-1. Specifications for the 5-volt power supply.

IC Design Projects

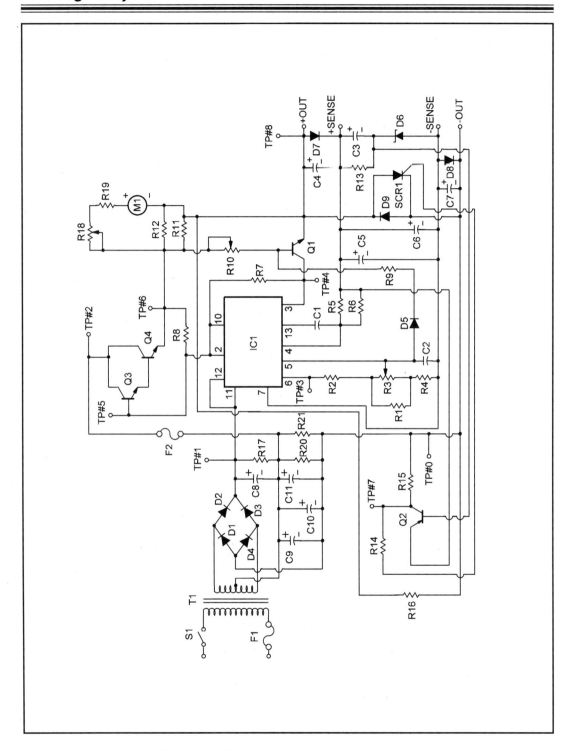

Figure 5-1. Schematic of the 5-volt power supply.

rectify the low AC voltage into a DC voltage. Bleeder resistors R17, R20 and R21 discharge filter capacitors C8 - C11 when the input voltage is removed.

Transistors Q3 and Q4 form a Darlington pair series-pass regulator. A Darlington pair provides a very high current gain. The series-pass regulator acts as an adjustable power resistor which changes in response to an error signal generated by the IC voltage regulator. The series-pass regulator changes its resistance to keep the output voltage constant under varying input voltage and load conditions. The series-pass regulator also reduces the ripple that is present on the input voltage.

Transistor Q1 monitors the voltage drop across the short-circuit current limit resistors, R11 and R12. The current flow through these resistors increases as the load current increases. When the load current becomes excessive, the voltage drop across R11 and R12 reaches 0.65 volts, and transistor Q1 conducts. When Q1 conducts, there is a voltage drop across resistor R7, which causes the current limiting transistor inside the LM723 to turn ON. This removes the drive from the series-pass regulator. The output current remains constant, and the output voltage is about 3.7 volts. Diode D5 starts to conduct and the foldback current limiting comes into play. The power supply acts as a constant current source.

When R10 is correctly adjusted, the output voltage increases when the short circuit is removed or decreased. Removing the short circuit decreases the voltage drop across resistors R11 and R12, which turns OFF transistor Q1 and allows for more base drive to the series-pass transistors, Q3 and Q4. The power supply acts as a constant voltage source.

Diodes D7 and D8 protect the power supply if the sense lines are opened. If the sense lines are opened, the output will be about 6.2 volts. If diodes D7 and D8 are missing, the output voltage of the power supply would be uncontrolled if the sense lines are opened.

Wire Gauge	Ohms/Ft.	Max. Length @ 2.5A	Max. Length @ 3.5A
22	0.0162	12.35 ft.	3.5 ft.
20	0.0101	19.8 ft.	5.6 ft.
18	0.00639	31.3 ft.	8.9 ft.
16	0.00402	49.75 ft.	14.2 ft.

Table 5-2. Wire gauge specifications.

IC Design Projects

The sense lines keep the load voltage constant regardless of the load current by compensating for any voltage drop across the wire connecting the load to the power supply. The sensing wires must be 22-gauge or larger to prevent errors due to the sense leads.

Table 5-2 can be used to calculate the maximum length of the sensing wires. The maximum length of each sensing wire is L = (voltage drop per lead)/(load current x wire resistance per feet).

Example 5-1

Use 16-gauge wire and a load current of 2.5 amperes. Calculate the maximum length of the power leads: L = 0.5/(2.5 x 0.00402) = 0.5/0.01005 = 49.75 feet.

Construction

The 5-volt power supply is an advanced project. The parts list for the power supply is listed in *Table 5-3*. This project can be built on a piece of perfboard or on a printed circuit board. You should use a low-wattage soldering iron with a small tip to build the power supply. Do not use a soldering gun because excessive heat can destroy transistors and integrated circuits. Keep the tip of the soldering iron clean by wiping the tip periodically with a damp rag or sponge. Always melt a new layer of solder onto the tip of the soldering iron each time the tip is wiped clean.

The transformer, rectifier and capacitor filter sections consisting of T1, D1 - D4, C8 - C11, R17, R20 and R21 should be built first. The subassembly should be verified before the rest of the power supply is built. With no load, and with TP#0 as the zero-volt reference, approximately 26 volts should be measured at TP#1, and approximately 13 volts should be measured at TP#2. When the subassembly is working properly, the rest of the power supply may be built.

Power transistor Q4 must be adequately heat sunk because it will dissipate a lot of heat in this project. Either the power transistor must be electrically isolated from the heat sink with a mica washer and bushings, or the heat sink must be electrically isolated from the power supply cabinet. A universal heat sink for transistors with a TO-3 case may be used. A TO-3 mounting socket and a mica insulator should be used to electrically isolate the power transistor from the heat sink.

Five-Volt Power Supply

All resistors are 1/4W @ 5% unless otherwise noted.	
All capacitors are 35V unless otherwise noted.	
R1, R4, R5:	3.9k
R2:	1k
R3, R10:	5k potentiometer
R6:	2.2k
R7:	510
R8, R15:	270
R9:	6.8k
R11, R12:	0.33 @ 5W
R13, R14:	100
R18:	10k potentiometer
R19:	10k
R20, R21:	1.2k @ 1/2W
Q1, Q3:	2N2484, 2N2222A
Q2:	2N2907A
Q4:	2N3055
IC1:	LM723
M1:	0-50uA
F1:	1A fast blow
F2:	4A fast blow
S1:	SPST
T1:	18 VCT secondary @ 3.5A
C1:	470 pF
C2:	0.047 uF
C3, C6:	1.0 uF tantalum
C4, C7, C8:	100 uF
C5:	470 uF
C9, C10, C11:	4700 uF
D1, D2, D3, D4:	1N5400
D5:	1N4148 or 1N5400
D6:	1N4735
D7, D8, D9:	1N4004
SCR1:	C122B

Table 5-3. Parts list for the 5-volt power supply.

You should also use an IC socket in the construction of this or any project. This way, the IC cannot be ruined by excessive heat, and can be easily replaced should it fail.

Testing and Calibration

Before applying power to the power supply, adjust potentiometer R10 fully clockwise. Jumper the +SENSE and the +OUT terminals together. Jumper the -SENSE and the -OUT terminals together. Connect a voltmeter to the +OUT and -OUT terminals. Apply power to the power supply and adjust potentiometer R3 for a 5 VDC (+/-0.1 volt) output. If all is well, turn off the power supply and proceed with the calibration procedure. If all is not well, verify the circuit operation using *Table 5-4*. Potentiometer R10 may not be properly adjusted. If the output voltage exceeds 6.5 VDC, check the wiring of zener diode D6, transistor Q2, and SCR1. If the output voltage is high but less than 6.5 VDC, verify the sense lead connections and the sense diodes. Potentiometer R3 may not be adjusted correctly.

Table 5-4 lists the proper direct voltage at various test points in the circuit. Do not proceed with the calibration until all of the test point voltages check out. Verify that the IC and all diodes, capacitors and transistors are properly oriented. Most problems are caused by solder bridges and cold solder joints.

With the power switch of the power supply OFF, connect a "DUMMY" load to the +OUT and -OUT terminals of the power supply. A suitable "DUMMY" load can be made with four 10-ohm, 10-watt resistors connected in parallel to form a 2.5-ohm, 40-watt load. Connect a voltmeter to the +SENSE and -SENSE terminals of the power

IC Design Projects

Test Point	No Load DC Volts	3A Load DC Volts
TP#1	26	22
TP#2	13	9.8
TP#3	7.2	7.2
TP#4	5.3	7.0
TP#5	5.3	7.0
TP#6	5.0	5.5
TP#7	0.0	0.0
TP#8	5.0	5.0

NOTE: TP#0 (-OUT terminal) is the zero volt reference point.

Table 5-4. Test point voltage.

supply. Apply power to the power supply. There should be 5 VDC across the load. If all is well, adjust potentiometer R18 for a two-ampere reading on the meter (meter M1 should indicate 20 uA). If the output voltage is zero or very low, check the wiring of transistor Q1.

Adjust potentiometer R10 fully counterclockwise. Connect the +OUT and -OUT terminals together. Both the output voltage and the output current should go to zero. When the connection between the +OUT and -OUT terminals is removed, the output voltage and current should stay at zero. This condition is known as a foldback latchup.

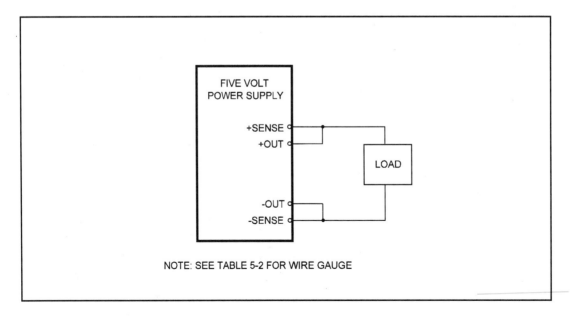

Figure 5-2. Connections for remote sensing.

Five-Volt Power Supply

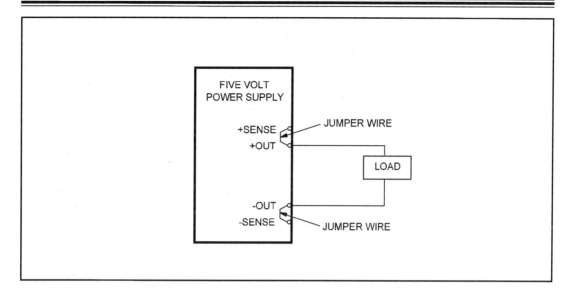

Figure 5-3. Connections when remote sensing is not required.

Connect the +OUT and -OUT terminals together. Adjust potentiometer R10 until the meter indicates 0.5 amperes (meter M1 should indicate 5 uA). Remove the connection between the +OUT and -OUT terminals. The output voltage should be 5 VDC and the output current should be two amperes (meter M1 should indicate 20 uA). Never adjust the foldback current for more than 1.5 amperes when the +OUT and -OUT terminals are connected together.

Using the Power Supply

The 5-volt power supply can be used to power transistor-transistor logic (TTL) circuits. Connect the +OUT terminal to the positive power rail of the circuit, and connect the -OUT terminal to the ground of the circuit to be powered by the power supply. Never power the load with only the +SENSE and -SENSE terminals. A voltmeter may be used to verify the output voltage. If necessary, potentiometer R3 may be used to adjust the output voltage of the power supply.

The remote sense terminals must be properly connected or the power supply will not function properly. If the load circuit is located far from the power supply, both the +OUT and +SENSE terminals must be connected to the positive power rail of the load circuit. Both the -OUT and -SENSE terminals must be connected to the ground of the load circuit. The proper connections are shown in *Figure 5-2*.

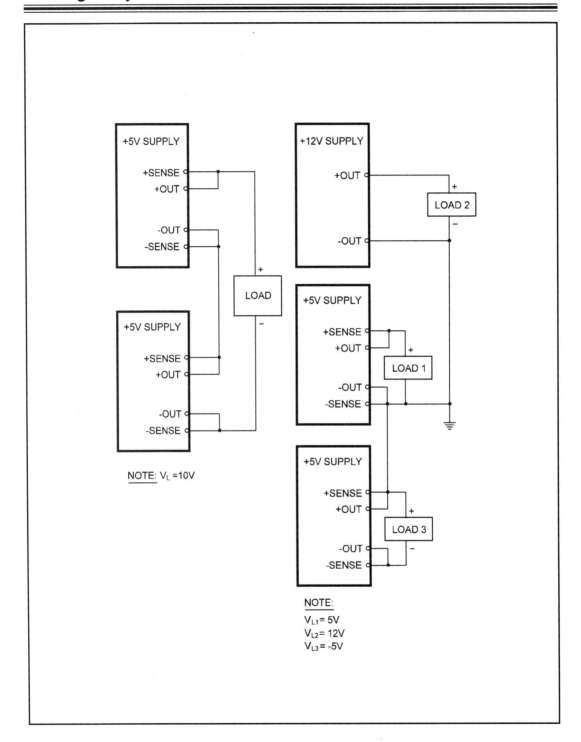

Figure 5-4. Connections for series operation of the power supply.

Five-Volt Power Supply

If the load is located near the power supply, the +SENSE and +OUT terminals must be connected together. The -SENSE and -OUT terminals must also be connected together as shown in *Figure 5-3*.

Two or more 5-volt power supplies may be connected in series because diode D9 protects the output electrolytic capacitor, C5, from sensing a reverse voltage greater than one volt, if one power supply is not turned on. The proper connections for series operation are shown in *Figure 5-4*. Parallel operation of two or more 5-volt power supplies is not recommended. It is better to divide the load circuit into two or more parts, and to drive each part with a separate 5-volt power supply.

Chapter 6
◆ Dual-Tracking ◆
◆ Power Supply ◆

Many electronic circuits require a variety of voltages of two polarities. This dual-tracking power supply fulfills these needs. The two-output voltages are adjustable from +/-1.2 volts to about +/-30 volts. If the TO-3 case LM317 and LM337 regulators are used, the power supply can deliver 1.5 amperes of current into a load. The dual-tracking power supply requires only readily-available components, and it is inexpensive to build.

Circuit Description

The schematic of the dual-tracking power supply is shown in *Figure 6-1*. Transformer T1 steps down the household AC line voltage to a more usable, lower AC voltage. Diodes D1 - D4 rectify the AC voltage into a pulsating DC voltage. Capacitors C1 and C2 filter some of the ripple component out of the pulsating DC voltage; the DC voltage is therefore converted to a fluctuating DC voltage. Capacitors C3 and C4 are signal

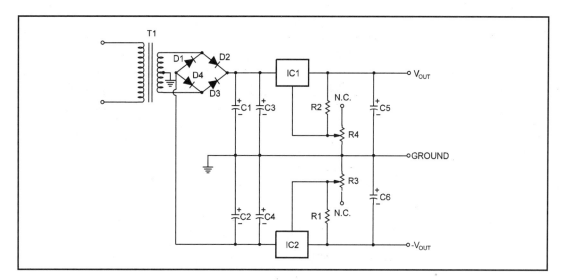

Figure 6-1. Schematic of the dual-tracking power supply.

```
All resistors are 1/4W @ 5% unless otherwise noted.

R1, R2:   220 ohms
R3, R4:   5k potentiometer
C1, C2:   2200 uF
C3, C4:   0.1 uF
C5, C6:   1.0 uF tantalum
D1-D4:    1N4004
IC1:      LM317
IC2:      LM337
T1:       50 VCT secondary @ 2A
```

Table 6-1. Parts list for the dual-tracking power supply.

bypass capacitors. Resistors R2 and R4 set the output voltage of the positive voltage regulator, IC1. Resistors R1 and R3 set the output voltage of the negative voltage regulator, IC2. Capacitors C5 and C6 improve the transient response of voltage regulators IC1 and IC2. The IC regulators, IC1 and IC2, remove the rest of the ripple component of the fluctuating DC voltages. The output voltages are therefore pure DC voltages.

Construction and Use

The dual-tracking power supply is easy to build. The parts list is shown in *Table 6-1*. The power supply may be built on a piece of perfboard. The IC regulators should be properly heat sunk by electrically isolating them from the heat sinks or the power supply cabinet. Suitable heat sinks can be made from two pieces of aluminum bent into shape. The dual-tracking power supply should work properly the first time it is powered up. It should provide years of reliable service.

Circuit Modifications

Transformer T1 can have any secondary voltage rating to a maximum of 56 VCT, because the maximum input voltage for the LM317 and LM337 regulators is 40 VDC. The output voltage of the LM317 and LM337 regulators is about 2.5 VDC less than their input DC voltage.

Resistors R1 and R2 can be replaced with 1% tolerance resistors. Potentiometers R3 and R4 can be substituted with a dual 5-kohm potentiometer. The output voltages would be simultaneously adjusted by the dual 5-kohm potentiometer. The output voltages will closely track each other if R1 and R2 are equal, and if the two halves of the dual potentiometer closely track each other.

Part Two
♦ TTL & CMOS ♦
♦ Logic Families ♦

Chapter 7
♦ Transistor-Transistor ♦
♦ Logic (TTL) ♦

TTL has been one of the most popular logic families for the past twenty-five years. TTL is available as ICs with a wide variety of logic functions. The 7400 series is the most popular line of TTL ICs. Only CMOS rivals TTL in popularity for use in digital systems employing SSI (small scale integration) and LSI (large-scale integration) packages. A typical TTL gate is shown in *Figure 7-1*.

The input circuit of the TTL gate consists of transistor Q1. Transistor Q2 is the driver stage, generating two complementary (out-of-phase) output signals. The complementary output signals drive the output circuit, which consists of transistors Q3 and Q4.

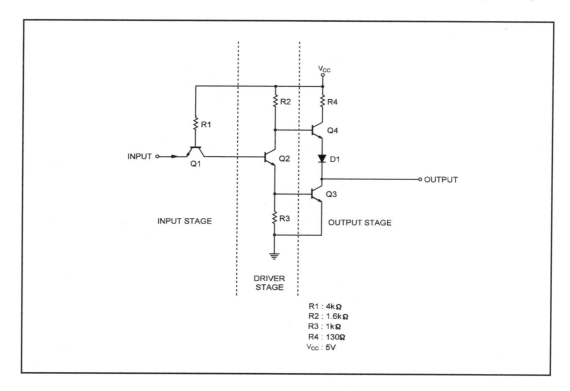

Figure 7-1. Schematic of a typical TTL gate.

IC Design Projects

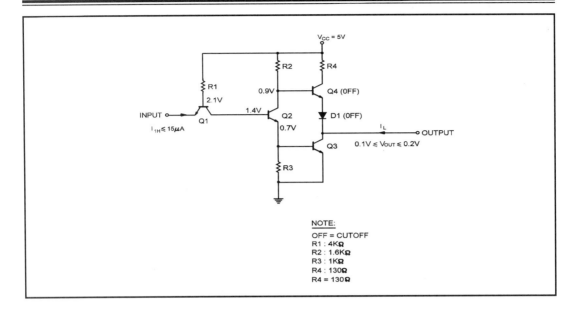

Figure 7-2. TTL gate voltage when the input is high.

Input Circuit Operation

If the input is high, current flows from Vcc through resistor R1. The base-collector junction of Q1 is forward-biased, and the base-emitter junction is reverse-biased. Transistor Q1 is operating in the inverse active mode; that is, in the active mode, but with the roles of the base and emitter interchanged. TTL gates have a very low reverse current gain. Therefore, TTL gates require a small input drive current. Transistor Q3 is driven into saturation and the output voltage of the gate is low (0.1V to 0.2V).

If the input is low, current flows from Vcc through resistor R4 and the emitter of Q1. The base-emitter junction of Q1 is forward-biased and the base-collector junction is reverse-biased. Transistor Q1 operates in its normal active mode. Transistor Q3 is driven into cutoff.

Output Circuit Operation

The output circuit consists of transistors Q3 and Q4. Transistor Q3 is a common-emitter circuit, and transistor Q4 is an emitter-follower circuit. Transistor Q3 can quickly discharge a capacitive load. Transistor Q4 can quickly charge a capacitive load.

Transistor-Transistor Logic (TTL)

In a steady state, Q3 provides a low resistance to ground; therefore, when the output is low, the gate can sink current through the saturated transistor Q3. Transistor Q4 provides the gate with a low current resistance, and it can therefore source current in the high state.

The output signal configuration is called a totem-pole output stage. Transistor Q4 is stacked on top of transistor Q3. Transistor Q4 is a pull-up transistor, because it pulls up the output voltage to the high level when the input being sent to the gate is low.

A special driver circuit is needed to generate the two complementary signals required by the totem-pole output circuit. The driver transistor, Q2, is a phase-splitter circuit because it provides two complementary signals.

Circuit Operation with Input High

The circuit voltages of a typical TTL gate, when the input is high, are shown in *Figure 7-2*. Transistor Q1 operates in the inverse active mode. The input current (I_{IH}) is very small, about 15 microamperes. The collector current of Q1 flows into the base of

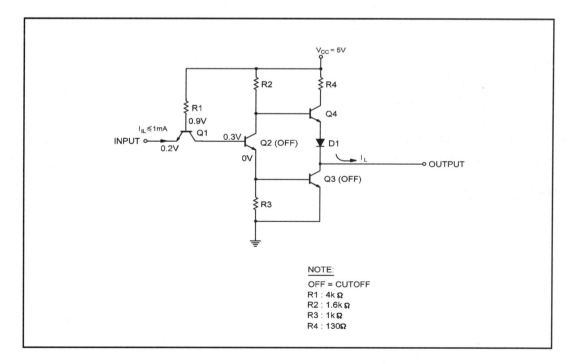

Figure 7-3. *TTL gate voltages when the input is low.*

IC Design Projects

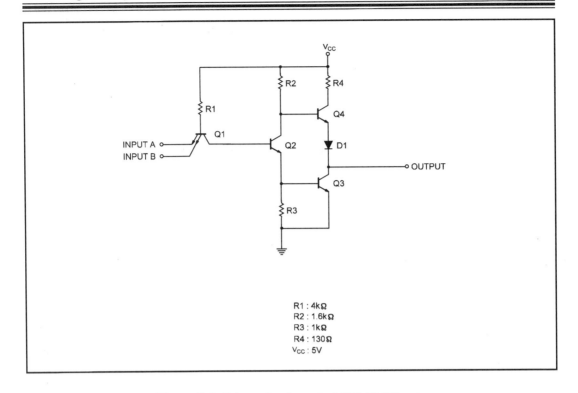

Figure 7-4. Schematic of a typical TTL NAND gate.

transistor Q2. The current is large enough to drive Q2 into saturation. The emitter current of transistor Q2 is sufficient to drive Q3 into saturation. The collector current of Q2 is enough to drive Q4 into saturation; however, for proper operation of the totem-pole circuit, transistor Q4 must be in the cutoff mode. Diode D1 ensures that transistor Q4 remains in its cutoff mode. The output voltage is low, approximately zero volts. In the low output state, the TTL gate can sink current into a load.

Circuit Operation with Input Low

The circuit voltages of a typical TTL gate, when the input is low (Vin = GND), are shown in *Figure 7-3*. Transistor Q1 operates in the saturation mode because, though its base-emitter junction is forward-biased, the base voltage (of Q1) of 0.9 volts is insufficient to forward-bias the series combination of the base-collector junction of Q1, and the base-emitter junction of Q2. Transistor Q2 remains in its cutoff mode. The gate input current, when the input is low (I_{IL}), is about one milliampere; it flows out of the emitter of Q1. If the TTL gate is driven by another TTL gate, output transis-

Transistor-Transistor Logic (TTL)

tor Q3 of the driving gate would wink I_{IL}. The maximum fan-out of a TTL gate is dependent upon I_{IL}.

If the TTL gate is not driven by another TTL gate, transistors Q2 and Q3 are in cutoff mode. Transistor Q4 is saturated, and it will supply or source current into a load. Transistor Q4 can either be in its active mode or in its saturation mode, depending upon the value of the load current.

If the load is an open circuit, transistor Q4 and diode D1 will be barely conducting because the load current is very small. If the load impedance is appreciable but small, the base-emitter junction of Q4 and diode D1 conduct more heavily, and Q4 is in its active mode. If the load draws a large current, transistor Q4 saturates. The output voltage is approximately equal to the power supply voltage.

The output voltage is determined by resistor R4. Resistor R4 limits the current flow through Q4, especially if the load is short-circuited. Resistor R4 also limits the supply current when transistor Q4 turns on while transistor Q3 is still in saturation.

TTL NAND Gate

The TTL gate shown in *Figure 7-1* is an inverter, because when the input is high, the output is low; and when the input is low, the output is high. The addition of a second base-emitter junction on transistor Q1, as shown in *Figure 7-4*, results in a two-input TTL NAND gate.

If either input (A or B) is low, the base-emitter junction of Q1 is forward-biased. There is no base drive for Q2; it is therefore in its cutoff mode. Transistor Q3 will also be in its cutoff mode. Depending upon the load current, Q4 is either in its active mode or saturation mode. The output of the NAND gate is high. When both inputs are high, the

INPUT	OUTPUT		INPUT A	INPUT B	OUTPUT
0	1		0	0	1
1	0		0	1	1
Inverter			1	0	1
			1	1	0
			NAND Gate		

Table 7-1. Truth tables for an inverter and a NAND gate.

NAND gate operates in a similar manner to the inverter gate and the output is low. When both inputs are low, the NAND gate operates in a similar manner to the inverter gate, and the output is high. The truth tables for the inverter and NAND gates are shown in *Table 7-1*. Two or more inputs to a gate are possible by fabricating the TTL integrated circuit with two or more base-emitter junctions on transistor Q1, making transistor Q1 a multi-emitter transistor.

TTL Specifications

The 7400 series TTL integrated circuits require a single power supply which must be 5 volts plus or minus half a volt.

The transfer characteristic of a typical TTL gate is shown in *Figure 7-5*. Segment AB indicates that when Q1 is saturated, transistors Q2 and Q3 are off, and Q4 is on. At

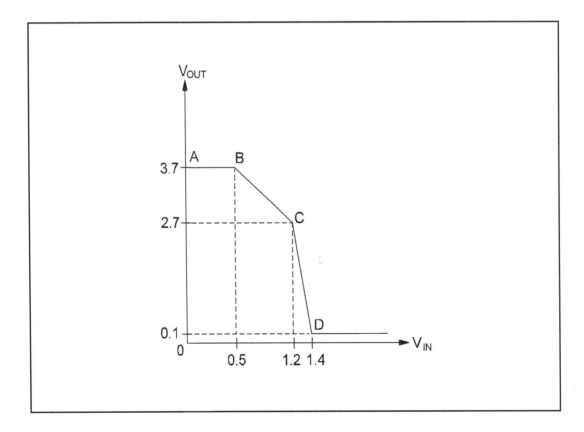

Figure 7-5. Voltage transfer characteristic of a TTL gate.

Transistor-Transistor Logic (TTL)

point B, transistor Q2 turns on because its base voltage reaches 0.6 volts. Segment BC shows when Q1 is still saturated, and Q2 turns on but operates as a linear amplifier. Diode D1 and transistor Q4 remain on and Q3 remains off. Breakpoint C corresponds to the moment when Q3 begins to conduct. Segment CD indicates when transistor Q3 operates in its active region. Transistor Q1 remains saturated, and Q2 and Q4 remain in their active regions. Breakpoint D corresponds to when transistors Q2 and Q3 saturate, and transistor Q4 is off.

TTL gates are immune to noise on the input signals. The zero level noise margin is ONM = V_{1L} - V_{0L} = 0.8 - 0.4 = 0.4 volts.

The one level noise margin is 1NM = V_{OH} - V_{1H} = 2.4 - 2.0 = 0.4 volts.

The propagation delay is the time between the 1.5 volts point of corresponding edges of the input and output waveforms. The propagation delay of a standard TTL gate is about 18 nanoseconds.

A TTL gate can dissipate five milliwatts when its output is high. It can dissipate 16.7 milliwatts when its output is low.

The fan-in of a TTL gate is 8, and the fan-out of a TTL gate is typically 12.

A TTL gate can switch at about 13 nanoseconds, which is the highest for saturated circuitry.

Other Forms of TTL Gates

The TTL logic family is versatile. Many different logic functions are available in SSI, MSI (medium scale integration) and LSI packages. There have been several improved versions of TTL gates developed over the years.

High-speed TTL ICs are available. Transistors Q1, Q2 and Q3 do not saturate in high-speed TTL gates. Their storage times are therefore reduced, resulting in higher switching speeds. The resistor values are also reduced in order to minimize the gate relay, by reducing the transistor time constants. High-speed TTL gates cannot dissipate as much power as regular TTL gates because the resistor values are reduced.

IC Design Projects

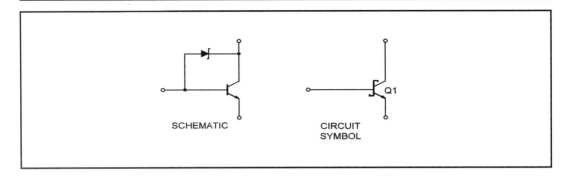

Figure 7-6. Schottky transistor.

Schottky TTL gates have very high switching speeds. The transistors are prevented from saturating by fabricating a Schottky diode across the base-collector junction of the transistor, as shown in *Figure 7-6*. A Schottky diode has a forward voltage drop of 0.4 volts. A Schottky transistor does not saturate because some of its base drive is shunted by the Schottky diode. The Schottky diode conducts and clamps the base-collector junction to a smaller voltage than that required for saturation. The storage time is eliminated, and the transistor turnoff time is reduced. Schottky diodes are added to transistor Q1, Q2 and Q3. Transistor Q4 does not require a Schottky diode because Q4 never saturates. Schottky TTL has a gate delay of about five nanoseconds, about half that of high-speed TTL.

Problems

Problem 7-1. What does TTL stand for?
Problem 7-2. What type of circuit can quickly discharge a capacitive load?
Problem 7-3. What type of circuit can quickly charge a capacitive load?
Problem 7-4. Which transistor can sink current from a load?
Problem 7-5. Which transistor can source current into a load?
Problem 7-6. What is the function of diode D1?
Problem 7-7. How can an inverter TTL gate be converted into a two-input NAND gate?
Problem 7-8. What power supply voltage is required for 7400 series TTL integrated circuits?
Problem 7-9. Define propagation delay.
Problem 7-10. What are the zero level and the one level noise margins of a typical TTL gate?
Problem 7-11. What is the propagation delay of a typical TTL gate?

Problem 7-12. What is the fan-in of a TTL gate?

Problem 7-13. What is the fan-out of a TTL gate?

Problem 7-14. How much power can a typical TTL gate dissipate?

Problem 7-15. How can the switching speed of a typical TTL gate be increased?

Problem 7-16. Can a high-speed TTL gate dissipate as much power as a regular TTL gate? Why or why not?

Problem 7-17. Can a Schottky transistor operate in its saturation region? Why or why not?

Chapter 8
◆ CMOS Logic ◆

Complementary metal-oxide silicon (CMOS) logic is popular in the design of medium-speed (up to 25 MHz) systems. CMOS offers excellent noise immunity and low power consumption, and can operate over a wide range of power supply voltages. CMOS logic is used in watches, clocks, automotive electronics, appliance controls and toys.

Lilienfeld first proposed the MOS field effect transistor in 1925. Heil proposed a similar structure in 1935. These early attempts at building a MOS transistor failed because of problems with the materials used. The bipolar transistor was developed in the Bell laboratories in 1948. The silicon planar process was developed in the early 1960s. MOS devices were finally introduced in 1967. CMOS technology is currently important as VLSI (very large scale integration) technology.

MOS Transistors

A metal oxide silicon (MOS) structure is fabricated by superimposing several layers of conducting, insulating and transistor forming materials.

A MOS transistor is a majority carrier device. The current in a conducting channel between the source and the drain is modulated by an input voltage applied to the gate. In an N-channel device or transistor, the majority carriers are electrons. A positive gate voltage, with respect to the substrate, enhances the number of electrons in the channel and therefore increases the conductivity of the channel. The channel is the region immediately under the gate. If the gate voltage is too small, the channel is cut off. The drain to source current is very low. The P-channel transistor operates in a similar manner, except that the majority carriers are holes and the gate voltage must be negative with respect to the substrate.

CMOS technology provides two types of transistors: the n-type or nMOS transistor, and the p-type of pMOS transistor. The n-transistor consists of a section of moderately doped p-type silicon separating two diffused areas of heavily doped n-type silicon. The gate of the n-transistor is an area separating the n-regions. The gate is a conducting electrode. The n-regions form the drain and the source of the n-transistor. Current flows from the drain to the source. The gate is the control input because it affects the flow of current from the drain to the source. The magnitude of the current flow is controlled by the gate-to-substrate potential difference or bias.

The p-transistor consists of a section of moderately doped n-type silicon separating two diffused areas of heavily doped p-type silicon. The gate of the p-transistor is an area separating the p-regions. The gate is a conducting electrode. The p-regions form the drain to the source of the p-transistor. Current flows from the drain to the source. The gate is the control input because it affects the flow of current from the drain to the source. The magnitude of the current flow is controlled by the gate-to-substrate potential difference or bias.

The MOSFET is biased such that the drain-to-substrate junction and the source-to-substrate junction are reverse-biased. The substrate is always the most positive voltage on the p-channel device. The substrate is always the most negative voltage on the n-channel device.

CMOS Inverter

The CMOS inverter consists of one n-channel and one p-channel MOSFET, as shown in *Figure 8-1*. The source of the p-device is connected to the positive terminal of the power supply. The source of the n-device is connected to the negative terminal or

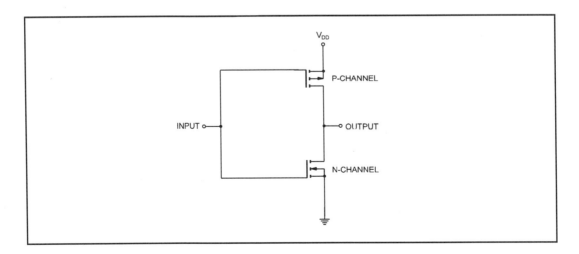

Figure 8-1. A CMOS inverter.

Complementary Metal-Oxide Silicon Logic

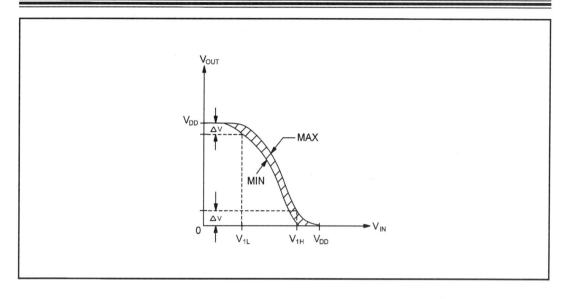

Figure 8-2. CMOS inverter transfer characteristic.

ground of the power supply. A single power supply in the range of 3 to 18 volts is required. Increasing the power supply voltage (within the 3-to-18-volts range) provides greater noise immunity and faster switching speeds at the expense of increased power dissipation.

When the inverter input is low or zero volts, the n-channel device is off and the p-channel device is on. The output voltage is high or equal to the positive power supply rail.

When the inverter input is high (power supply voltage), the n-channel device is on and the p-channel device is off. The output voltage is low or zero volts.

The transfer characteristic of the CMOS inverter is shown in *Figure 8-2*. The CMOS transfer characteristic is nearly ideal with regard to noise immunity, because the CMOS zero level is very close to zero volts, and the CMOS one level is very close to the power supply positive voltage. If the n-channel and p-channel devices are matched, the switching threshold voltage is exactly one-half of the power supply voltage. The gain at the switching point is usually high. There is usually a mismatch between the *n* and *p* devices in practical CMOS circuits. Therefore, switching does not occur at exactly one-half of the power supply voltage. Manufacturers of CMOS ICs usually specify minimum and maximum transfer characteristics, as shown in *Figure 8-2*.

IC Design Projects

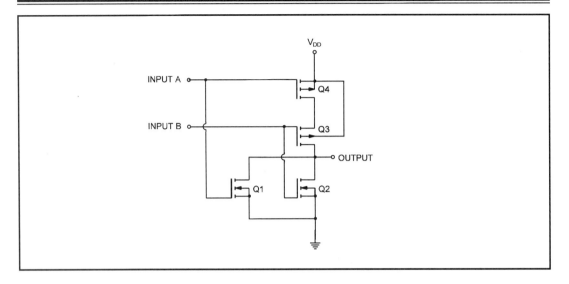

Figure 8-3. Two-input CMOS NOR gate.

CMOS NOR and NAND Gates

A two-input CMOS NOR gate is shown in *Figure 8-3*. Transistors Q3 and Q4 are p-channel devices connected in series and to the positive rail of the power supply. Transistors Q1 and Q2 are n-channel devices, connected in parallel and to the ground potential of the power supply. The output voltage is high only when both inputs, A and B, are low. The truth table for the NOR gate is listed in *Table 8-1*.

A two-input CMOS NAND gate is shown in *Figure 8-4*. Transistors Q1 and Q2 are p-channel devices, connected in parallel and to the positive rail of the power supply. Transistors Q3 and Q4 are n-channel devices, connected in series and to the ground potential of the power supply. The output voltage is high when input A, input B, or both inputs are low. The truth table for the NAND gate is listed in *Table 8-1*.

Specifications

CMOS ICs require a power supply voltage in the range of 3 to 18 volts to operate. The majority of CMOS logic circuits are powered by a DC voltage in the range of 5 to 15 volts.

Complementary Metal-Oxide Silicon Logic

Noise margin is the allowable noise voltage on the input of a gate, so that the output of the gate is not affected. The worst-case noise margins for CMOS logic are:

$V_{DD} = 5V$, $0NM = 1NM = 0.5V$

$V_{DD} = 10V$, $0NM = 1NM = 1V$

$V_{DD} = 15V$, $0NM = 1NM = 1V$

A CMOS gate dissipates several nanowatts of power when it is not in the switching mode. When a CMOS gate IS in the switching mode, it dissipates several milliwatts of power. The power dissipated is proportional to the frequency of operation of the CMOS gate. A typical CMOS gate dissipates 6 milliwatts of power at a switching frequency of one megahertz, and with a power supply voltage of 10 volts.

PMOS transistors have a slower switching sweep than nMOS transistors, by a factor of ten. CMOS devices incorporate pMOS and nMOS transistors for a reasonable switching speed. CMOS devices have a slower switching speed than TTL devices.

The propagation delay of a CMOS gate is greater than that of a TTL gate. The switching speed of a CMOS gate is less than that of a TTL gate. CMOS gates have a high fan-in and a high fan-out.

INPUT A	INPUT B	OUTPUT
0	0	1
0	1	0
1	0	0
1	1	0

NOR Gate

INPUT A	INPUT B	OUTPUT
0	0	1
0	1	1
1	0	1
1	1	0

NAND Gate

Table 8-1. Truth tables for NOR and NAND gates.

IC Design Projects

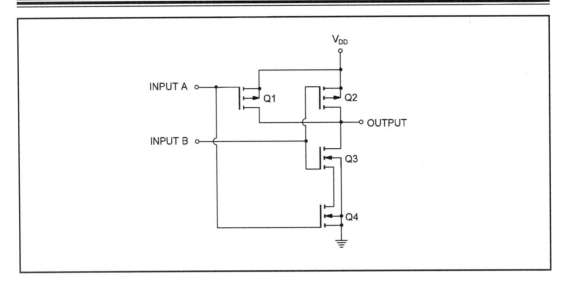

Figure 8-4. Two-input CMOS NAND gate.

The gate delay of a CMOS device is fairly constant. As the supply voltage increases, the current available to charge and discharge load capacitances also increases. The output voltage swing also increases.

CMOS gates are slightly more expensive than TTL gates.

Proper Handling of CMOS Integrated Circuits

The input resistance of MOS devices is very high. Static electricity can cause charge to accumulate on the input capacitance during handling. The charge can generate large voltages, which in turn can cause the device to break down. The oxide insulator between the metal gate and the substrate of a CMOS device breaks down at a gate to substrate potential of about 100 volts. The IC is destroyed if the oxide insulator breaks down even once. CMOS devices have input diodes that limit the input voltage to a safe level.

An unconnected gate input can float at an unknown voltage because the input resistance of a CMOS device is high. Leakage currents are usually sufficient for the input device to enter its switching mode, allowing large currents to flow and causing the device to overheat. Unused inputs should be connected to either the positive power supply rail or to ground.

As with TTL gates, CMOS gates cannot have its output terminals connected together to create wired-AND or wired-OR logic. Tristate output devices must be used.

CMOS integrated circuits must be handled with care. Unused CMOS devices should be stored in conductive foam. Alternatively, the leads can be shorted with aluminum foil. The soldering iron should have a grounded tip. Use grounded assembly tables, if possible. Always touch the ground with one hand and pick up the CMOS device with the other hand. Never exceed the maximum voltage, current and temperature ratings of the CMOS device.

CMOS logic is close to being the ideal logic family. CMOS is currently available in SSI, MSI, LSI and VLSI packages. LVSI packages include memory and microprocessor integrated circuits.

Problems

Problem 8-1. What does CMOS stand for?
Problem 8-2. What are some advantages to using CMOS ICs?
Problem 8-3. How is a MOS transistor fabricated?
Problem 8-4. How does a MOS transistor function?
Problem 8-5. What is the gate of a MOS transistor?
Problem 8-6. What controls the current flow from the drain to the course of a MOS transistor?
Problem 8-7. How should a MOSFET be biased?
Problem 8-8. What happens to a CMOS inverter when the power supply voltage is increased within the 3- to 18-volt range?
Problem 8-9. If the N-channel and P-channel devices are matched, what is the switching threshold voltage?
Problem 8-10. Why do manufacturers specify minimum and maximum transfer characteristics?
Problem 8-11. What is the required power supply voltage for CMOS circuits?
Problem 8-12. Define *noise margin*.
Problem 8-13. What are the noise margin specifications for CMOS integrated circuits?
Problem 8-14. How much power does a CMOS gate dissipate?
Problem 8-15. What are some precautions necessary when handling CMOS integrated circuits?

Chapter 9
◆ Capacitance Meter ◆

This capacitance meter autoranges and measures capacitance from 1 pF to 1 uF, and from 1 uF to 4000 uF. It updates readings automatically. The capacitance meter is useful for determining the values of unmarked capacitors.

Capacitors are usually measured on an AC bridge by balancing the reactance of known components against the reactance of an unknown capacitor at a fixed frequency. The meter measures capacitance by measuring time. Transistor-transistor logic (TTL) circuits are used in the capacitance meter because they have a high switching speed.

The unknown capacitor is first charged to a known voltage. It is then discharged to another known voltage through a fixed resistance. The discharge time is directly proportional to the unknown capacitance, and is determined by the capacitance meter.

Circuit Description

The schematic of the power supply of the capacitance meter is shown in *Figure 9-1*. Transformer T1 steps down the household line voltage to 12.6 volts center-tapped. Diodes D1 and D2 rectify the low AC voltage to a pulsating DC voltage. The positive regulator IC, U1, provides 5-volts DC to the rest of the capacitance meter circuits. Capacitor C1 smooths out the pulsating DC voltage, and C2 improves the transient response of U1. The rest of the capacitance meter circuit is shown in *Figure 9-2*.

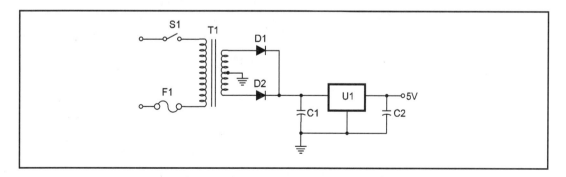

Figure 9-1. Power supply for the capacitance meter.

IC Design Projects

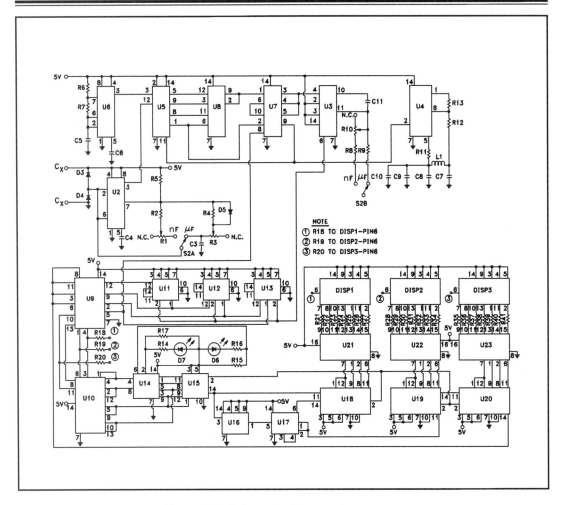

Figure 9-2. Schematic of a capacitance meter.

The unknown capacitance becomes the timing capacitor for U2, which is configured as an astable multivibrator. When switch S2 is in the nF position, components R1, R2 and Cx determine the discharge time of Cx. The discharge time of Cx is determined by components R3, R4 and Cx when S2 is in the uF position.

A second LM555 timer IC, U6, is also configured as an astable multivibrator. It is part of an autocycling circuit, which automatically updates the capacitance measurement. The autocycling circuit is supplied with a 1.4 MHz reference clock. The clock is generated by integrated circuit U4, which is configured as a Colpitts oscillator. Components L1 and C7 - C10 form the tank circuit for the Colpitts oscillator.

All resistors are 1/4W @ 5% unless otherwise noted
R1: 100k trimmer potentiometer
R2: 1M ohm @ 1%
R3: 100 ohm trimmer potentiometer
R4: 1 kohm @ 1%
R5: 1 kohm
R6-R7: 100 kohms
R8-R9: 1.5 kohms
R10: 25 kohm linear potentiometer
R11-R13: 100 ohms
R14-R15: 3.3 kohms
R16-R41: 470 ohms
D1-D2: 1N4002
D3-D5: 1N4154
D6-D7: LED
L1: 12 uH
S1: SPST
S2: DPDT
F1: 1/4A fast blow
T1: 12.6 V.C.T. secondary

All capacitors rated at 16 volts
C1: 4700 uF
C2: 0.1 uF
C3: 0.0033 uF
C4, C6, C7: 0.01 uF
C5: 4.7 uF
C8: 820 pF
C9: 470 pF
C10: 220 pF
C11: 0.005 uF
U1: LM7805
U2, U6: LM555 timer
U3, U16: 74121
U4, U14: 7404
U5, U8: 7474
U7, U10, U17: 7400
U9: 74125
U11-U13, U18-U20: 7490
U15: 7593
DISP1-DISP3: DL707 or common-anode display

Table 9-1. Parts list for the capacitance meter.

The outputs of the astable multivibrators and the Colpitts oscillator are combined by U5 and U8, which are dual-D flip-flops. One-half of U5 synchronizes the output of U2, with the output of the Colpitts oscillator providing dual-phase outputs. The other half of U5, and integrated circuit U8, select one discharge pulse from U2 when the output of the autocycler astable multivibrator, U6, is high. Astable multivibrator U6 is disabled by the flip-flops until the discharge pulse is completed.

The Colpitts oscillator output is gated by U7, then it passes to the counting stages during one discharge period of each measuring interval of the unknown capacitor. Monostable multivibrator U3 resets the decade counters, U15 and U18 - U20, and dividers U11 - U13. Monostable multivibrator U3 is triggered by the leading edge of the synchronized discharge pulse. When switch S2 is in the nF position, the reset pulse width of U3 is controlled by the zero-trimmer potentiometer, R10. This eliminates the effect of stray capacitance on the measurement.

The gated Colpitts oscillator output is divided by decade counters U11 - U13. The counter outputs are fed to the tristate logic switch, U9, which passes the pulse train to decade counter U20. Overflow pulses generated by counter U20 are passed on to decade counters U19 and U18. The BCD outputs of the decade counters are decoded by U21 - U23, which are BCD to seven-segment decoder/drivers. Integrated circuits U21 - U23 drive the seven-segment displays, DISP1 - DISP3. Resistors R21 - R41 limit the current to each segment of each display. Resistors R18 - R20 limit the current to the decimal points of DISP1 - DISP3.

Overflow pulses from the last decade counter U18 are applied to the four-bit binary counter, U15. The inverted outputs of U14 are decoded by U10 (thus providing control signals to the tristate logic switches) and U9. They select the proper display decimal point, as well as sink or block current from overrange indicators D6 and D7.

Construction and Verification

The capacitance meter is an advanced project, as can be seen from the parts list of *Table 9-1*. The meter can be built on a piece of perfboard. This project can also be wirewrapped.

The capacitance meter can be tested after it is completely built. You should make sure that this project is built and tested in stages.

The power supply should be the first section that is built and tested. The power supply circuit is shown in *Figure 9-1*. If you are using a transformer without a center tap, diodes D1 and D2 should be replaced by the diode rectifier bridge consisting of diodes D1, D2, D8 and D9, as shown in *Figure 9-3*. Once the 5-volt power supply is built and working, the circuits of *Figure 9-2* can be built and verified.

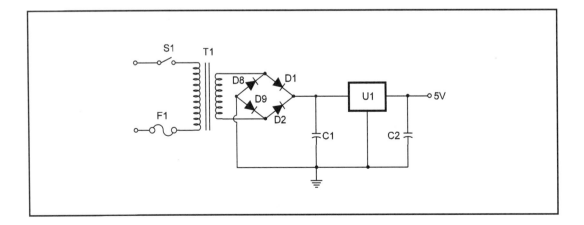

Figure 9-3. Alternate power supply for the capacitance meter.

Capacitance Meter

When building the astable multivibrator circuit of U2, a square wave should be present at U2 pin 3 when a capacitor is connected to the Cx terminals. Verify that this stage works for both settings of switch S2.

The oscillator, synchronizer and reset circuits of U3 - U8 should be built next. Do not install capacitors C9 and C10 yet. A square wave should be present at U6 pin 3 and at U4 pin 2. There should be a square wave at U7 pin 8 and at U3 pin 6. There is also a square wave at U5 pin 12, as long as Cx is connected to astable multivibrator U2.

The autoranging circuit is built next. It consists of integrated circuits U9 - U20 and associated components. There should be square waves or pulses at U11 pins 1 and 12, at U12 pins 1 and 12, as well as at U13 pins 1 and 12. The overflow light-emitting diodes should illuminate only in the overflow condition.

The display circuits should be built last, and consist of BCD to seven-segment decoder/drivers U21 - U23, DISP1 - DISP3 and associated components. When there is no capacitor connected to the Cx terminals, the display should indicate a "000" readout within a couple of seconds, if potentiometers R1, R3 and R10 are rotated counterclockwise and S2 is in the nF position.

Calibration and Use

When the zero potentiometer R10 is rotated clockwise, the display should indicate a few picofarads. Rotate R10 until the display indicates "000." Connect a capacitor known to be 0.68 uF to the meter. Set switch S2 to the uF position, and adjust potentiometer R3 until the display indicates "0.68." Set S2 to the nF position and adjust R1 until the display indicates "680." If the required display is not obtainable, install capacitors C9 and C10, and repeat the calibration procedure.

Each time the capacitance meter is used with S2 in the nF position, the display should be "zeroed" with the zero potentiometer. When S2 is in the uF position, no zeroing is required. The unknown capacitor is connected to the Cx terminals, with the positive capacitor lead connected to the positive Cx terminal. The unknown capacitor must be discharged to protect the input circuitry of the capacitance meter.

If the unknown capacitance is more than 1000 nF, place S2 in the uF position. If the unknown capacitance is less than 1000 nF, place S2 in the nF position. If the unknown capacitance is greater than 1000 uF, place S2 in the uF position. The capacitance can be determined by observing the overrange LEDs. If only the upper LED glows, the unknown capacitance is 1000 uf. If only the lower LED glows, the unknown capacitance is 2000 uF. If both LEDs glow, the unknown capacitance is 3000 uF. If the sequence repeats, then the unknown capacitance is 4000 uF or greater, depending on how often and what part of the sequence repeats.

Capacitors with high leakage will not charge to the reference voltage, and therefore will not trigger the discharge cycle. When S2 is in the nF position, and the extreme left decimal point is lit, treat the display as reading the unknown capacitance in pFs. A display of ".047" should be read as 47 pF.

Chapter 10
◆ Digital Logic Probe ◆

A digital logic probe is used to measure logic levels of digital circuits. It indicates the logic level by illuminating the appropriate LED (light-emitting diode). This digital logic probe can indicate low levels, high levels and pulse streams. It also has a one-bit memory to store the previous logic level measurement.

The digital logic probe is powered by the "circuit under test." The logic probe is useful for testing TTL and CMOS circuits. A measuring device should present as light a load

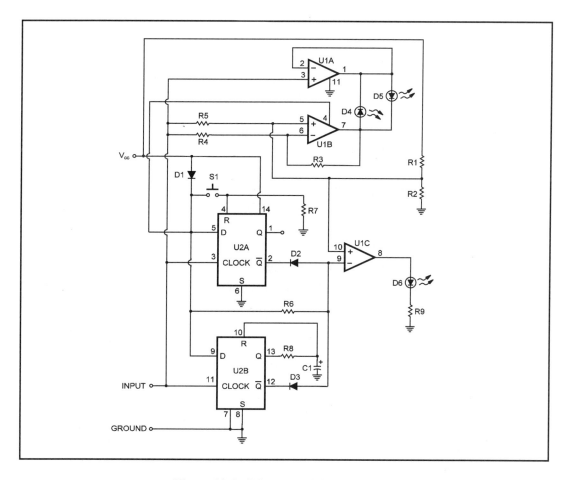

Figure 10-1. *Schematic of the logic probe.*

IC Design Projects

as possible to the circuit under test, to ensure that the measuring device is indicating true readings of the circuit. This digital logic probe uses a CMOS integrated circuit because CMOS ICs consume very little power.

Circuit Description

The digital logic probe is a simple circuit, as can be seen from the schematic shown in *Figure 10-1*. The digital logic probe is powered by the power supply of the circuit under test.

The CD4013 is a dual-D flip-flop. Each flip-flop has its own D (data), R (reset), S (set) and clock inputs, as well as Q and NOT Q outputs. The D input is transferred to the Q output on the leading or positive-going edge of the clock pulse only when the inputs R and S are low. The D flip-flops of the CD4013 may be set or reset by applying a high level to the S or R input, respectively.

The LM324 is an IC consisting of four independent operational amplifiers. The LM324 IC can therefore replace four operational amplifier ICs. The common-mode input range includes the negative power supply, thus eliminating the need for external biasing components. The LM324 requires only a single-ended power supply in the range of 3 volts to 32 volts.

Op-amp U1A is configured as a unity-gain non-inverting amplifier. Op-amp U1B is configured as a unity-gain inverting amplifier. Op-amp U1C is configured as a high-gain inverting amplifier.

The logic probe interprets its input as low when the input voltage is less than the reference voltage. The reference voltage is set by resistors R1 and R2. When a low level is applied to the input of the logic probe, U1 pin 7 goes high because the input to U1 pin 6 is less than the reference voltage at U1 pin 5. At the same time, U1 pin 1 goes low. Light-emitting diode D4 is forward-biased, and therefore illuminates to indicate that a low-level signal is applied to the input of the logic probe. Light-emitting diode D5 remains dark because it is reverse-biased. The low level input does not affect the D flip-flop outputs. Light-emitting diode D6 serves as the memory output device. Diode D6 remains dark.

The one-bit memory consists of U1C, integrated circuit U2, and the associated support components. The memory is reset by momentarily pressing switch S1.

All resistors are 1/4W at 5% unless otherwise noted	
R1:	12k
R2:	5.6k
R3-R5:	100k
R6, R7:	22k
R8:	15k
R9:	120 ohms
C1:	10 uF at 25 volts, tantalum
D1-D3:	1N914 or 1N4148
D4-D6:	Light-emitting diode
U1:	LM324 quad operational amplifier
U2:	CD4013 dual D flip-flop
S1:	Normally-open push-button switch

Table 10-1. Parts list for the digital logic probe.

The logic probe interprets its input as a high when the input voltage is greater than the reference voltage. The reference voltage is set by resistors R1 and R2. When a high level is applied to the input of the logic probe, U1 pin 7 goes low because the input to U1 pin 6 is greater than the reference voltage at U1 pin 5. At the same time, U1 pin 1 goes high. Light-emitting diode D5 is forward-biased, and therefore illuminates to indicate that a high-level signal is applied to the input of the logic probe. Light-emitting diode D4 remains dark because it is reverse-biased. The high level input forces the NOT Q outputs of the D flip-flops low, forward-biasing diodes D2 and D3. Light-emitting diode D6 illuminates because U1 pin 8 is high. The memory can be reset by momentarily pressing switch S1.

When a pulse is applied to the input of the logic probe, light-emitting diodes D4 and D5 alternately illuminate if the input pulse frequency is less than about 12 kilohertz. Diodes D4 and D5 remain constantly illuminated at higher input pulse frequencies. The NOT Q outputs of U2A and U2B alternate between the low and high levels. Light-emitting diode D6 lights when the NOT Q outputs are low because U1 pin 8 is high when the NOT Q outputs are low. If the reset switch, S1, is pressed, diode D6 alternately illuminates and extinguishes, or remains illuminated, depending on the frequency of the pulse stream that has been detected.

The logic probe shows 0% to 20% Vcc as a low logic level. It shows 40% to 100% Vcc as a high logic level. The trigger level of the logic probe is approximately 56% Vcc. TTL and CMOS level detection are possible. It can capture pulse widths of less than 250 nanoseconds. Diode D1 protects the digital logic probe in the event that the power leads are connected in reverse.

IC Design Projects

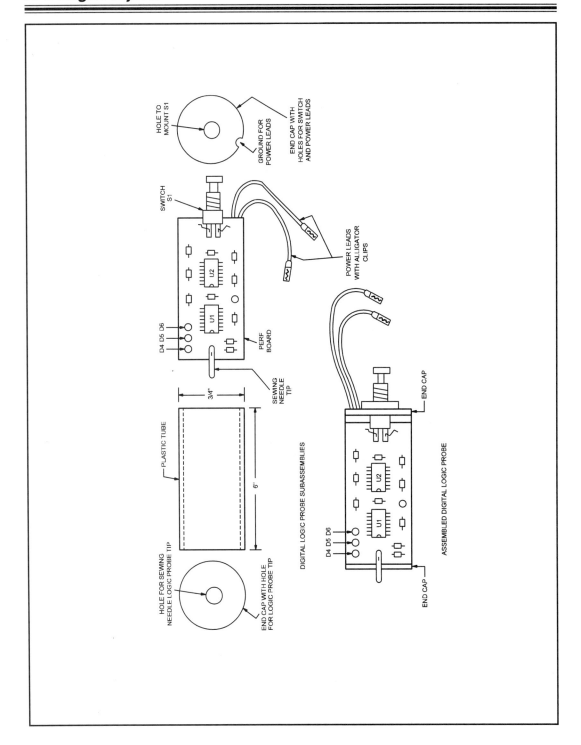

Figure 10-2. Cabinet construction details.

Page 100

Construction

The digital logic probe is easy to build. It requires only two integrated circuits and several support components. The parts list for the logic probe is given in *Table 10-1*.

The logic probe can be built on a piece of perfboard. Take care to properly orient the integrated circuits, the diodes and the capacitor. Use IC sockets to prevent soldering iron heat damage to the integrated circuits.

A low-wattage soldering iron should be used to prevent heat damage to the components. Take care to avoid cold solder joints and solder bridges. Most operating problems are caused by poor soldering techniques, cold solder joints and short circuits created by solder bridges. The inexperienced hobbyist should practice soldering components onto a scrap piece of perfboard.

The logic probe can be housed in any type of cabinet. The housing can be a 6" length of a transparent plastic tube with an inside diameter of 1/3." In this case, the perfboard should measure 5-1/2" by 3/4." The cabinet construction details are shown in *Figure 10-2*.

The IC sockets should be installed first. The resistors, diodes and capacitors are installed next. Switch S1 can be mounted on an end cap, which can be a piece of alumi

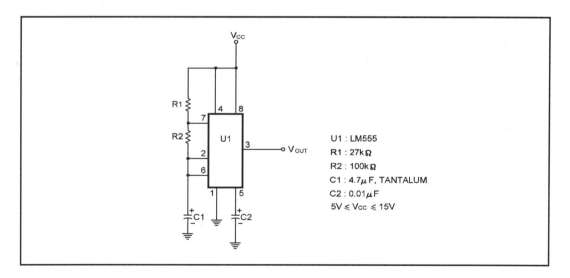

Figure 10-3. Schematic of the test circuit.

IC Design Projects

num cut to size. Then wire the switch into the circuit. The logic probe tip can be a sewing needle. The plating on the "eye" end of the needle may have to be sanded off before you solder the needle into the circuit. The power leads should be about twelve inches in length, each with an alligator clip soldered to one end. The power leads should be color coded; red for the positive power lead, and black (or green) for the ground lead. The power leads are passed through a groove in an end plate, as shown in *Figure 10-2*. Finally, the integrated circuits are installed into the IC sockets. The final assembly of the digital logic probe in its cabinet, as shown in *Figure 10-2*, should not be done until the logic probe is tested and working properly.

A second end plate is made from a piece of aluminum, as shown in *Figure 10-2*. Drill a small hole so that the sewing needle logic probe tip can pass through the end cap.

Testing and Use

The digital logic probe can be tested by wiring the test circuit shown in *Figure 10-3*. The test circuit is an astable multivibrator with a square-wave output whose frequency is about one hertz, with a duty cycle of forty-four percent. The resistor and capacitor values are not critical. The digital logic probe and test circuit should be powered by the same power supply. The power supply voltage should be in the range of 5 volts to 15 volts.

Touch the digital logic probe tip to the positive power supply rail. Light-emitting diodes D5 and D6 should illuminate. When the tip is removed from the power supply positive rail, diode D5 extinguishes and diode D6 remains illuminated. Pressing reset switch S1 should extinguish light-emitting diode (LED) D6.

Touch the digital logic probe tip to the ground of the power supply. Only LED D4 should illuminate. When the tip is removed from the ground, LED D4 extinguishes.

Touch the digital logic probe tip to the output of the test circuit (U1 pin 3). LEDs D4 and D5 should either alternately illuminate and extinguish, or diodes D4 and D5 should remain illuminated, depending on the output frequency of the test circuit. When reset switch S1 is kept pressed, diode D6 should pulse on and off, or illuminate constantly, depending on the frequency of the pulse stream detected.

Once the digital logic probe is fully functional, final assembly in its case should be carried out, as shown in *Figure 10-2*. Once this is completed, the digital logic probe is ready for use.

The digital logic probe is easy to use. You only have to touch the logic probe tip to the point of the circuit being tested, and the device does the rest.

A single-pulse or a multiple-pulse burst can be latched in by pressing the reset switch momentarily, while holding the logic probe tip on the test point of the circuit under test. The first pulse will activate the probe memory, and LED D6 illuminates until it is reset by pressing the reset switch, S1.

A continuous pulse train can be detected by keeping the reset switch S1 pressed. LED D6 either pulses on and off, or continuously illuminates, depending on the frequency of the pulse train. Frequencies above 12 kilohertz can be detected by keeping the reset switch S1 pressed. LED D6 illuminates continuously.

Part Three
♦ Operational ♦
♦ Amplifiers ♦

Chapter 11
◆ The Operational ◆
◆ Amplifier ◆

The first operational amplifier (op-amp) appeared about fifty years ago. They were expensive, heavy and fragile tube amplifiers. The tube operational amplifier was used to perform mathematical operations in analog computers. The operational amplifier gets this designation because it was originally used to perform mathematical operations on electrical signals. The operational amplifier is currently available as an integrated circuit for less than one dollar.

Early op-amps were built using transistors and resistors. The first IC op-amp (uA7-9) was introduced around 1965. Its performance characteristics were poor by today's standards, and it was quite expensive. Electronic engineers started using op-amps in their designs; consequently, the performance characteristics of op-amps improved while their prices dropped.

The op-amp is a linear IC building block. The op-amp can be used for a wide variety of linear and nonlinear electronic circuits, depending upon the feedback network connected between its input and output terminals. Multiple-stage transistor amplifier circuits with their bias design requirements are a thing of the past.

The Ideal Operational Amplifier

The op-amp is essentially a high-gain differential amplifier. A differential amplifier has two input terminals and one output terminal. The output voltage is proportional to the difference between the two input voltages.

The typical operational amplifier consists of a differential amplifier input stage, a level-shifter stage, and an output power amplifier stage. The input differential amplifier may be followed by one or more differential amplifiers, if additional gain is required. The level shifter drives the output power amplifier. The various stages are directly coupled. No coupling capacitors are used because a constant differential gain is required down to very low frequencies, including DC. The differential amplifier input stage provides high input impedances, as well as a high differential gain. The level-shifter stage provides voltage gain. The output power amplifier stage provides voltage gain and a low output impedance.

Operational amplifiers are complex circuits using several transistors and resistors. The ideal op-amp has infinite gain, is direct coupled, has an infinite impedance and a zero output impedance, and has two differential inputs. Differential inputs permit op-amps to function with balanced (or ungrounded) signals, providing flexibility in processing signals.

The circuit of an ideal op-amp is shown in *Figure 11-1*. The output voltage of an op-amp is equal to the product of its differential gain (or open-loop gain), A, and the difference between its two input voltages. The open-loop gain of an op-amp is the gain without a feedback loop: $V_{OUT} = A(V_{IN2} - V_{IN1})$.

The ideal op-amp does not draw any input current because the input impedances are infinite. The output terminal of an op-amp acts like an ideal voltage source because the output impedance is zero. The output voltage is therefore independent of the current drawn from the output terminal into a load impedance.

The output voltage is in phase with the noninverting (+) input voltage, and it is out of phase with the inverting input (-) voltage. The output voltage is dependent on the difference of the input voltages. There is no output voltage when the input voltages are equal.

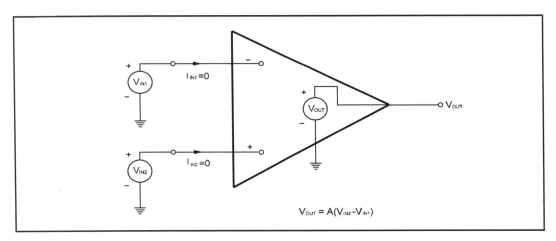

Figure 11-1. Equivalent circuit of the ideal op-amp.

The ideal op-amp has an infinite bandwidth. The open-loop gain remains constant from zero (or DC) frequency to infinite frequency.

The ideal op-amp also has an infinite open-loop gain. In most applications, the op-amp is not used in its open-loop configuration. A feedback network is usually required to close the loop around the op-amp. The feedback network can be either a negative feedback network or a positive feedback network.

The Practical Operational Amplifier

Operational amplifiers have a very high open-loop gain, in the range of several thousand to several hundred thousand. The typical op-amp has an open-loop gain of 200,000. The higher the open-loop gain, the better.

Op-amps have a high input impedance. Op-amps that have bipolar transistors on the input differential amplifier stage can have an input impedance of about 1,000,000 ohms. An input impedance of several megohms is possible if field-effect transistors (FET) are used for the input differential amplifier stage. The higher the input impedance, the better.

Op-amps have a low output impedance, usually in the range of 50 ohms to 300 ohms. The typical op-amp has an output impedance of 75 ohms. When a feedback loop is connected to the op-amp, the output impedance can be less than one ohm. The lower the output impedance, the better. The high input impedance and the low output impedance of the op-amp allows it to be cascaded without using buffer stages.

The op-amp requires its output to be fed back to its input through a feedback network, consisting of resistors or capacitors. When the output is fed back to the inverting input through a feedback network, the desired frequency and gain responses are obtained with the negative feedback op-amp circuit. An op-amp will oscillate if a positive feedback loop is placed between its output and input terminals. An op-amp may be internally or externally compensated to prevent unwanted oscillations.

Op-amps usually require dual (or bipolar) power supplies because op-amps incorporate several differential amplifiers in cascade to provide high gain and common-mode rejection. Common-mode rejection is the ability of an op-amp to cancel out common signals fed to its inverting and noninverting inputs. The two inputs allow phase inversion for negative feedback, and they can be connected to provide in-phase or out-of-phase amplification.

The practical op-amp has a small mismatch between the transistors of the differential amplifier input stage. A small DC voltage, or output offset voltage, will be present at the output of the op-amp when there is no input voltage (the input terminals are grounded) to the op-amp. The output offset voltage may be reduced by applying a small DC differential voltage to the inputs of the op-amp. The output offset voltage can also be affected by temperature changes.

The base bias currents for the transistors of the input differential amplifier stage require a relatively low impedance path. Coupling capacitors on the input circuit are therefore objectionable, unless an alternate low impedance DC path is provided for each input terminal. The input offset current is the difference of the two input base bias currents when the output DC voltage is zero. The input bias current is the average value of the two base bias currents when the output DC voltage is zero.

Temperature changes affect the input offset current of the op-amp and the input bias current of the op-amp.

Specifications

Most op-amps require dual or bipolar power supplies. Op-amps generally require power supply voltages in the range of +/-9 volts to +/-15 volts.

The supply current is the quiescent current required by the op-amp under no-load conditions. Low current drain is advantageous in battery-powered circuits. Op-amps usually require less than 5 milliamperes of supply current.

Offset voltage is the output voltage that is generated by the op-amp when there is no input voltage applied to the op-amp. The output offset voltage is due to the input bias current and the input offset voltage. Low offset voltage is mandatory in direct-coupled circuits. The offset voltage is usually less than 100 millivolts.

The input bias current is the current that must be supplied to each input of an op-amp for proper biasing of the input differential amplifier stage transistors. The input bias current can be less than one nanoampere, or as high as 2 microamperes.

The Operational Amplifier

The input offset voltage is the differential input voltage that must be applied to the op-amp for a zero output voltage. The input offset voltage is usually 10 millivolts or less.

Op-amps have a high input impedance of at least one megohm. Op-amps have a low output impedance of approximately 100 ohms. These impedances allow for the cascading of op-amp stages.

The input offset current is the difference between the input bias currents flowing into each op-amp input when the output voltage is zero. The input offset voltage is less than 500 nanoamperes.

Slew rate is the maximum rate of change of the op-amp output voltage in response to a square-wave differential-mode input signal. Slew rate is critical in high-speed and low-distortion circuits. The slew rate is usually less than 10 volts per microsecond.

The common-mode rejection radio (CMRR) is the ability of an op-amp to cancel out (within the device) common signals fed to its inverting and noninverting inputs. CMRR is important when designing summing and differential amplifiers. The CMRR of an op-amp is at least 60 decibels.

The power supply voltage rejection ratio (PSRR) is the ability of an op-amp to prevent power supply voltage fluctuations from appearing in the output voltage. PSRR is important when designing battery-powered circuits, and in circuits where offset voltages must be kept as low as possible. The PSRR of an op-amp is less than 200 microvolts per volt.

Most electronic components generate noise. Low noise is essential in high-quality audio and video circuits. The typical op-amp generates a low noise voltage, less than 100 nanovolts per square root of the frequency of the input signal.

The gain of an op-amp is its amplification factor. The gain of an op-amp is usually at least fifty thousand.

The gain of an op-amp decreases as the frequency of its input signal increases. The gain-bandwidth product of an op-amp is constant for all frequencies, ensuring a flat and linear operation. The frequency at which an op-amp has unity gain is its unity-gain bandwidth. The typical op-amp has a unity-gain bandwidth in the range of one megahertz to twenty megahertz. The IC op-amp is not suitable for high-frequency applications.

IC Design Projects

The maximum specification for the power supply voltage of an op-amp must never be exceeded. The collector-base junctions of the IC transistors will break down, causing large currents to flow. The IC op-amp will overheat, causing permanent damage.

The input signal peak-to-peak amplitude should never exceed two or three volts. Excessive input voltages can destroy the base-emitter junctions of the input stage transistors of the op-amp.

The power supply voltages must never be connected in reverse. The normally reverse-biased isolation junctions of the op-amp will become forward-biased. Large currents will flow through the isolation junctions, destroying the op-amp.

The output stage of an op-amp can typically dissipate one half of a watt. If the op-amp does not incorporate short-circuit protection in its output stage, an accidentally short-circuited load can destroy the IC op-amp.

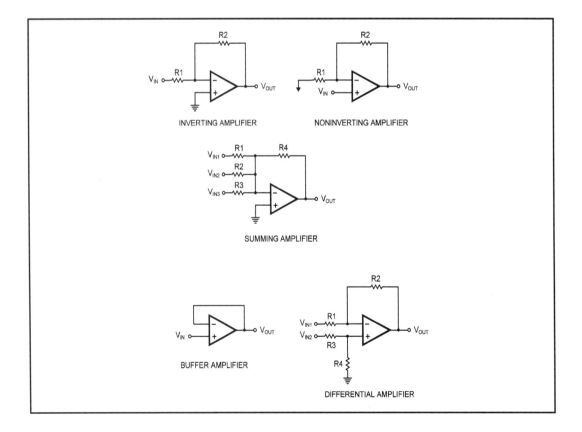

Figure 11-2. Amplifier circuits using op-amps.

Operational Amplifier Circuits

External components can be connected to the op-amp to modify its characteristics, improve its performance, or customize its function. If the external components are connected between the output and the inverting input terminals, negative feedback is obtained. If the external components are connected between the output and the noninverting input terminals, positive feedback is obtained.

An op-amp is operating in its linear range when the output voltage is directly proportional to its input voltage. When the op-amp is in saturation (that is, the output reaches its maximum or minimum excursion), the op-amp is operating in its nonlinear range.

Linear Circuits

The inverting amplifier is shown in *Figure 11-2*. A resistor is connected from the output to the inverting input. A second resistor accepts the input signal. The noninverting input terminal is connected to ground. The output signal is out of phase with the input signal. The minus sign indicates out of phase: $V_{OUT} = -R2(V_{IN})/R1$ and $R_{IN} = R1$.

The noninverting amplifier has a resistor connected between the output and the inverting input terminals, as shown in *Figure 11-2*. The inverting input terminal is connected to ground via a second resistor. The input signal is fed to the noninverting terminal. The input and output signals are in phase. The output resistance is very small, a few milliohms: $V_{OUT} = V_{IN}(1 + R2/R1)$ and $R_{IN} = AR1/(1 + R2/R1)$. A is the open-loop gain.

The summing amplifier is also shown in *Figure 11-2*. A resistor is connected between the output and the inverting input terminal. The inputs are connected to the inverting terminal by other resistors. The noninverting input terminal is connected to ground: $V_{OUT} = -V_{IN1}(R4/R1) - V_{IN2}(R4/R2) - V_{IN3}(R4/R3)$. $R_{IN} = 1/(G1 + G2 + G3)$, where $G_x = 1/R_x$.

The buffer or voltage-follower amplifier is also shown in *Figure 11-2*. The output and inverting input terminals are connected together. The input signal is fed to the noninverting input terminal. This circuit has a very high input impedance and a very low output impedance. It has a unity gain: $V_{OUT} = V_{IN}$. $R_{IN} = AR'_{IN}$, where R'_{IN} is the op-amp open-loop input impedance. $R_{OUT} = R'_{OUT}/A$, where R'_{OUT} is the op-amp open-loop output impedance.

IC Design Projects

The differential amplifier is shown in *Figure 11-2*. Signals are applied to both inputs of the op-amp. The output signal is the difference between the two input signals: $V_{OUT} = (R4/(R3 + R4))((R1 + R2)/R1)V_{IN2} - (R2/R1)V_{IN1}$ and $R_{IN} = R1 + R3$.

In mathematics, integration and differentiation are inverse operations. An op-amp can be used to perform integration and differentiation operations.

An op-amp integrator is shown in *Figure 11-3*. A capacitor is connected between the output and inverting input terminals. The input voltage is fed through a resistor to the inverting input terminal. The noninverting input terminal is connected to ground. The output signal is the inverted integral of the input signal. An integrator produces a triangular output waveform when a square wave is fed to its input: $V_{OUT} = -1/R1C1\{Integral(V_{IN})dt\}$, where R1C1 is the period of the signal to be integrated.

An op-amp differentiator is shown in *Figure 11-3*. A resistor is connected between the output and inverting input terminals. The input is fed to the inverting input terminal

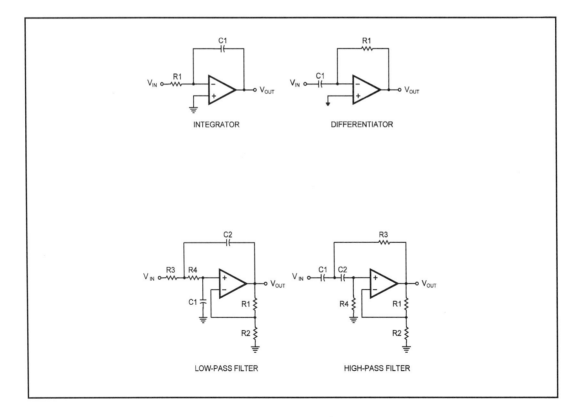

Figure 11-3. Other linear op-amp circuits.

through a coupling capacitor. The noninverting input terminal is connected to ground. A square-wave output waveform is generated when the input is a triangular waveform: $V_{OUT} = -R1C1\{d(INPUT)/d(TIME)\}$, where R1C1 is the period of the signal to be differentiated.

There are dozens of other linear op-amp circuits, including low-pass, high-pass, notch and bandpass filters. Sallen and key low-pass and high-pass filters are shown in *Figure 11-3*. A low-pass filter only passes frequencies below the cutoff frequency. A high-pass filter only passes frequencies above the cutoff frequency. A bandpass filter only passes the frequencies between its two cutoff frequencies. A notch filter passes all frequencies except those between its two cutoff frequencies.

Nonlinear Circuits

Integrated circuit op-amps can be used for digital applications. Linear and nonlinear op-amp circuits may be cascaded, and the same power supply can be used for both circuits.

A comparator circuit is shown in *Figure 11-4*. There is no feedback component. The comparator is an open-loop circuit. The inverting input is connected to a reference voltage. If the input voltage is greater than the reference voltage, the output voltage is equal to the positive supply voltage. If the input voltage is less than the reference voltage, the output voltage is equal to the negative supply voltage: $V_{OUT} = A(V_{IN} - V_{REF})$.

The bistable multivibrator (or flip-flop) has two stable states. In the op-amp circuits shown in *Figure 11-4*, the two stable states are positive and negative saturation. An input trigger signal is required to change the output from one stable state to the other stable state. A DC-coupled flip-flop circuit is shown in *Figure 11-4*, which also shows an AC-coupled flip-flop circuit.

In the astable multivibrator shown in *Figure 11-4*, a square wave is generated at the output as the multivibrator circuit switches back and forth between its two stable states. The astable multivibrator is also called a free-running oscillator.

In the monostable multivibrator shown in *Figure 11-4*, the op-amp is normally in positive saturation. When a positive trigger pulse is applied to the input, the output swings into negative saturation. The monostable multivibrator automatically returns to its positive saturation state after a time interval. The pulse width, T, depends on compo-

IC Design Projects

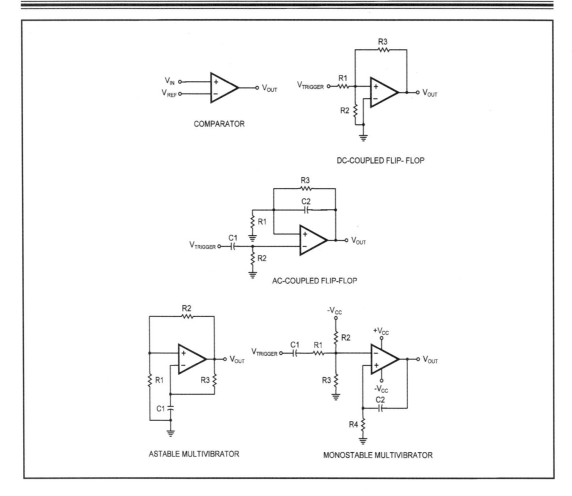

Figure 11-4. Nonlinear op-amp circuits.

nents R4 and C2. The monostable multivibrator is also called a one-shot multivibrator: T = -R4C2($V_{REF}/V_{FEEDBACK}$).

There are dozens of other nonlinear op-amp circuits that you might want to study in volumes outside of this one, including log and antilog amplifiers, precision rectifiers and voltage regulators.

Problems

Problem 11-1. How did the op-amp get its name?
Problem 11-2. What are the stages of an op-amp?

The Operational Amplifier

Problem 11-3. Why can an op-amp operate down to DC?

Problem 11-4. What are the ideal characteristics of an op-amp?

Problem 11-5. Define open-loop gain.

Problem 11-6. What is the output voltage of an op-amp when the two input voltages are equal?

Problem 11-7. What are the typical specifications of a practical op-amp?

Problem 11-8. What type of power supply do most op-amps require?

Problem 11-9. Define offset voltage.

Problem 11-10. Define input offset voltage.

Problem 11-11. Define slew rate.

Problem 11-12. Define common-mode rejection ratio.

Problem 11-13. If an op-amp has a gain-bandwidth product of 1,000,000 Hz, at what frequency is its gain equal to ten?

Problem 11-14. How can an op-amp be destroyed?

Problem 11-15. When is an op-amp operating in its linear range?

Problem 11-16. When is an op-amp operating in its nonlinear range?

Problem 11-17. In an inverting amplifier, R2 = 10 kohms, R1 = 1 kohm, and V_{IN} = 0.1 volts. Calculate V_{OUT} and R_{IN}.

Problem 11-18. In a noninverting amplifier, R2 = 10 kohms, R1 = 1 kohm, A = 100,000 and V_{IN} = 0.1 volts. Calculate V_{OUT} and R_{IN}.

Problem 11-19. What is the V_{OUT} of a summing amplifier if R1 = R2 = R3 = R4?

Problem 11-20. What is the V_{OUT} of a differential amplifier if R1 = R3 and R2 = R4?

Problem 11-21. What is the output of an op-amp integrator if the input signal is a cosine wave?

Problem 11-22. What is the period of the signal to be differentiated if R1 = 1 kohm and C1 = 0.01 uF?

Problem 11-23. Define low-pass, high-pass, bandpass and notch filters.

Problem 11-24. When is the output of an op-amp comparator equal to the positive supply voltage?

Problem 11-25. What is the pulse width of an op-amp one-shot multivibrator if R4 = 1 kohm, C2 = 0.01 uF, V_{REF} = 10 volts and $V_{FEEDBACK}$ = 1 volt?

Chapter 12
♦ Home Theater System ♦

A home theater system is a surround-sound system. Surround sound is the reproduction of the spacious acoustics of a live performance in a small listening room. This home theater system consists of two buffered input stages, a surround-sound decoder, a rear channel amplifier and two front channel amplifiers.

Surround sound in residential environments is possible because of the introductions of stereo television and the stereo video cassette recorder (VCR).

Most households have a stereo system in which the left front speaker reproduces the left side of the live performance, and the right front speaker reproduces the right side of the live performance. There are no rear speakers in a stereo system.

Dolby Laboratories invented a system of encoding the rear channel information on the existing stereo channels. A decoder is required to decode the rear channel signal and

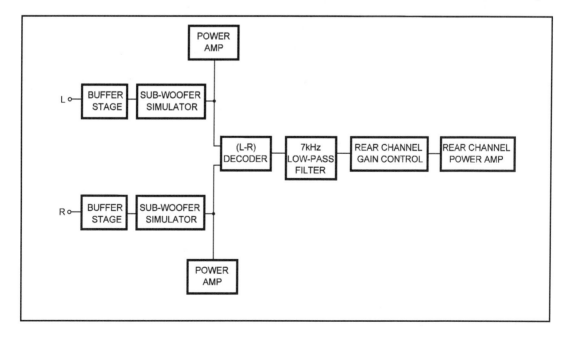

Figure 12-1. Block diagram of the home theater system.

IC Design Projects

feed it to the rear channel amplifier. Most surround-sound decoders are designed for use in the movie theater. The home theater system in this book was designed by the author for use in residential environments. A block diagram of the system is shown in *Figure 12-1*.

The most popular surround-sound system is the Dolby system. The rear channel signal is generated by subtracting the right channel signal from the left channel signal. Therefore, an L-R decoder is required for the home theater system. The decoder passes the left input signal in phase and inverts the right input signal. High frequencies are attenuated by a low-pass filter. The low-pass filter passes all frequencies in its passband in phase with its input signal. The rear channel amplitude is carried by the rear channel gain control. The rear channel signal is amplified by the rear power amplifier and sent to the rear channel speakers.

The average residential listening room is about twelve feet wide. The front speakers are placed on either side of the stereo television. Speech will therefore appear to emanate from the television. An L + R or monophonic center voice channel is not required in a home theater system.

Sound effects such as explosions or a closing door contain a lot of low-frequency information. Most stereo systems do not respond to subaudio frequencies, and would therefore benefit from a subwoofer simulator.

The subwoofer simulator boosts the low-frequency response of the system. The simulator consists of an input amplifier, a low-pass filter, and a summing amplifier. *Figure 12-2* is a block diagram of the subwoofer simulator.

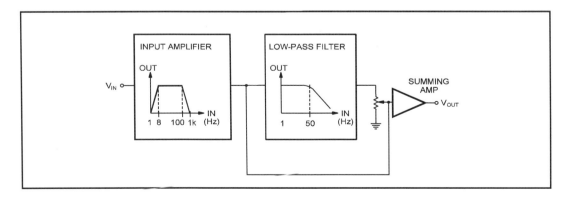

Figure 12-2. Block diagram of subwoofer simulator.

Home Theater System

Circuit Operation

The left and right channel signal sources are selected by switch S1 and fed to integrated circuits U1 and U3, which serve as buffer amplifiers. Integrated circuits U1 and U3 are configured as voltage followers. The voltage follower is a noninverting amplifier with unity gain, a high input impedance and a low output impedance. The LM741 operational amplifier is selected for U1 and U3 because it is internally compensated, has high gain, and is not subject to latchups.

Sound effects such as explosions or thunder contain a lot of information at frequencies below 50 Hz. Most amplifiers roll off frequencies below 30 Hz; therefore, very low frequencies must be amplified, which requires a subwoofer simulator incorporated into

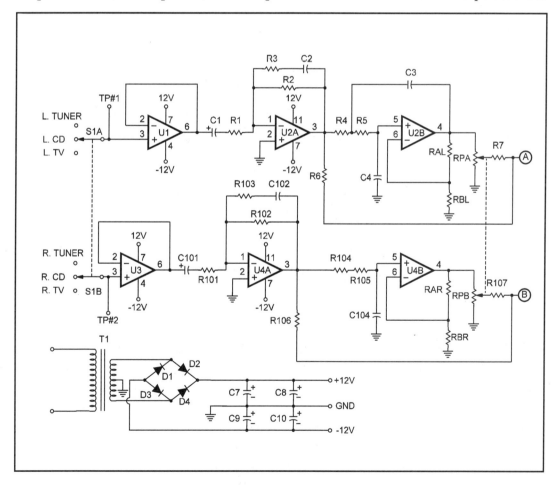

Figure 12-3a. Schematic of the home theater system.

IC Design Projects

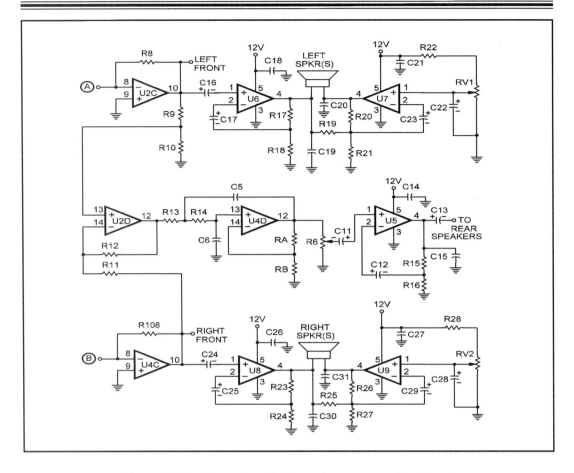

Figure 12-3b. Schematic of the home theater system (continued).

each front channel. This enables the amplifiers to reproduce the information present at low frequencies.

The schematic of the home theater system and its power supply is shown in *Figure 12-3*. U2A, U2B and U2C form the left channel subwoofer simulator, while U4A, U4B and U4C form the right channel subwoofer simulator. The subwoofer simulator provides unity gain at frequencies above 1 kHz, providing up to 28 dB of gain at 10 Hz.

Capacitors C1 and C101 are coupling capacitors, used to block DC signals from entering the subwoofer simulator. At very low frequencies, C1, C101, C2 and C102 act as open circuits. The gain of the input amplifiers are therefore less than unity, at very low frequencies. At very high frequencies, C1, C101, C2 and C102 act as short circuits.

Home Theater System

The gain of the input amplifiers are therefore unity at higher frequencies. The bass boost is adjustable via dual potentiometer RP.

The decoder derives an L-R rear channel signal from the two front channel signals. High frequencies are attenuated by a low-pass filter. The rear-channel amplitude is varied by the rear gain control, RG. The decoder passes the left input signal in phase and inverts the right input signal.

The low-pass filter blocks high-frequency signals. High-frequency sounds are directional. In live performances, high-frequency sounds are heard only when the listener is on an axis with the high-frequency source.

U2D is the L-R decoder, which is a unity gain differential amplifier where the left channel signal is fed to the noninverting input of the operational amplifier, and the right channel is fed into the inverting input of the operational amplifier. U4D is configured as a 7 kHz low-pass filter. The rear-channel gain control is potentiometer RG.

The rear channel power amplifier can be any amplifier capable of driving two 8-ohm speakers in parallel. The LM383 (U5 of *Figure 12-3*) is a cost-effective power amplifier suitable for audio applications. High current capability (3.5A) enables the device to drive low impedance loads with low distortion. The LM383 is current-limited and thermally protected. The LM383 in a 5-pin TO-220 package is shown in *Figure 12-4*. The LM383 is pin-for-pin compatible with the TDA2002. The maximum power sup-

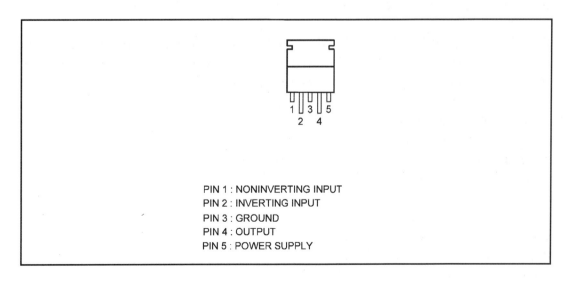

Figure 12-4. Pin configuration of the LM383 IC.

IC Design Projects

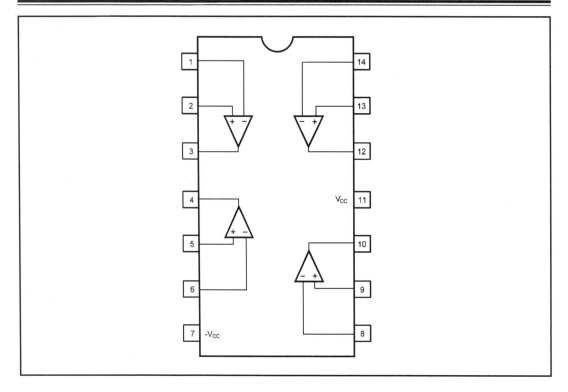

Figure 12-5. Pin diagram of the RC4136 IC.

ply voltage allowed for the LM383 IC is 14 VDC. In *Figure 12-3*, the LM383 is used as the rear channel power amplifier, U5.

The LM383 can deliver eight watts when used as a basic power amplifier, as shown in *Figure 12-3*. The LM383 can deliver sixteen watts when used as a bridge amplifier, also shown in *Figure 12-3*, where it serves as the left and right channel amplifiers.

Two LM383 integrated circuits are required for the bridge power amplifier. The left channel bridge power amplifier consists of ICs U6 and U7, and associated components. Potentiometer RV1 serves as the left channel volume control. The right channel bridge power amplifier consists of ICs U8 and U9, and associated components. Potentiometer RV2 serves as the right channel volume control.

The remaining components in *Figure 12-3* make up the bipolar power supply required to power the system. Transformer T1 steps down the household line voltage to 18 VCT AC voltage. Diodes D1 - D4 rectify the low AC voltage to a pulsating DC voltage. Capacitors C7 - C10 smooth the pulsating DC voltage to a low ripple DC voltage.

```
All resistors are ¼ watt @ 5% unless otherwise noted.
All capacitors are rated 25 volts.

R1, R101:  47 kohms
R2, R102:  270 kohms
R3, R103:  56 kohms
R4, R5, R104, R105:  33 kohms
R6, R8-R12, R106, R108:  10 kohms
R7, R107, RA, RAL, RAR:  1800 ohms
RB, RBL, RBR:  3000 ohms
RP, RV1, RV2:  100-kohm dual potentiometer
R13, R14:  22 kohms
RG:  5000-ohm potentiometer
R15, R17, R19, R20, R23, R25, R26:  220 ohms
R16, R18, R21, R24, R27:  10 ohms
R22, R28:  1.0 megohms
D1-D4:  1N5408
C1, C101:  1.0 uF
C2, C102:  0.047 uF
C3, C4, C103, C104:  0.1 uF
C5, C6:  0.001 uF
C7-C10:  4700 uF
C11, C16, C22, C24, C28:  10 uF
C12, C17, C23, C25, C29:  470 uF
C14, C15, C18-C21, C26, C27, C30, C31:  0.22 uF
C13:  2000 uF
T1:  18 V.C.T. secondary
U1, U3:  LM741
U2, U4:  RC4136
U5-U9:  LM383 or TDA2002
S1:  2P3T rotary
```

Table 12-1. Parts list for the home theater system.

Construction

The surround-sound decoder may be built on a piece of perfboard using point-to-point wiring. *Figure 12-5* is a diagram of the pin configuration of the RC4136 quad op-amp IC being used. Each RC4136 contains four separate op-amps. Alternately, eight LM741 op-amps may be used as long as the proper pins are used for the inverting and noninverting inputs, the output and power supply leads.

The surround-sound decoder is a moderately complex project, as can be seen from the parts list in *Table 12-1*. Care should be taken to orient the capacitors, diodes and ICs properly, and to avoid cold solder joints and short circuits created by solder bridges. Most operational problems are caused by cold solder joints and solder bridges.

IC Design Projects

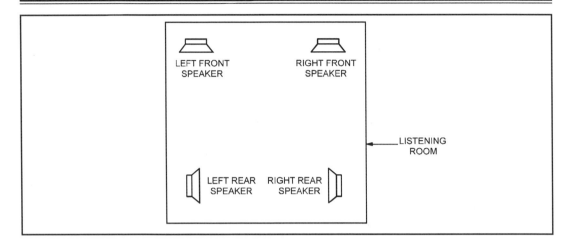

Figure 12-6. Proper rear speaker placement.

Testing

Testing the decoder is straightforward. Connect the same signal to TP#1 and TP#2. There should be a signal at the output of U2C and U4C. There shouldn't be a signal at the output of U2D; hence, no sound from the rear speakers. If either TP#1 or TP#2 is grounded, and a signal is fed to the other input, there should be a signal present at the output of U2D, and therefore from the rear speakers.

These initial tests should be conducted using a 1-kHz sine wave for the test signal. Signals lower than 100 Hz will be amplified and can be adjusted using the bass control, RP. The gain of the rear channel can be adjusted with gain control RG. The frequency response of the rear channel should roll off above 7 kHz.

Installation

The placement of the rear speakers is important. They should be mounted at the rear of the side walls about six feet above the floor, as shown in *Figure 12-6*. If the speakers are mounted on the rear wall, they will not have ambient-reflected sound. Proper placement should bring exciting, three-dimensional surround sound into your listening room.

Chapter 13
◆ Alpha Brain-Wave ◆
◆ Feedback Monitor ◆

Biopotentials are very small voltages present in all living organisms. In human beings, biopotentials are caused by the activity of nervous system sensors, muscles and nerves. All biopotentials are generated by living cells.

The electrocardiogram (EKG) is a recording of the electrical activity of the heart. The electroencephalogram (EEG) is a recording of the electrical activity of the brain. The electromyogram (EMG) is a recording of the electrical activity of the muscles.

Depending upon the mental activity, brain waves occur in the frequency range of 0.2 hertz to 28 hertz. Brain wave characteristics are listed in *Table 13-1*. Alpha brain waves occur at 7.5 hertz to 13 hertz. They occur when a person is at rest or in a state of heightened awareness. Theta brain waves occur when a person is solving problems, planning for the future, or daydreaming. Delta brain waves occur in the frequency range of 0.2 hertz to 3.5 hertz. They occur when a person is asleep. Beta brain waves, in the frequency range of 13 to 28 Hz, occur when a person is worried, surprised, angry or hungry.

The brain is a voltage source, represented electrically in *Figure 13-1*. The brain generates a common-mode signal, Ec, though a common-mode source impedance, Zc. The brain is a balanced voltage source; therefore, there are two source impedances, Zs1 and

Name of Brain Wave	Magnitude	Frequency	Mental State	Percent
Alpha	10-100 uV	7.5-13	Tranquillity, relaxation, heightened awareness.	10%/day
Theta	50-200 uV	3.5-7.5	Uncertainty, problem-solving, future planning, day dreaming, switching thoughts.	25%/day
Delta	10-50 uV	0.2-3.5	Deep sleep, trance sleep, non-REM sleep.	10%/day
Beta	10-50 uV	13-28	Worry, anger, fear, attention, tension, hunger, surprise.	35%/day
Other	0.01-0.1 uV	Vhf-Uhf	Unknown	Unknown

Table 13-1. *Brain-wave characteristics.*

IC Design Projects

Zs2. The ratio, Zs1/Zs2, can vary between one and 100 or more. The ratio can therefore affect the common-mode rejection ratio of the differential input circuit of the brain-wave monitor. The common-mode signal includes all unwanted signals such as electrode potentials, power line interference, and noise from extraneous body signals. The large fields can be screened out by a differential amplifier. A differential amplifier rejects signals common to both inputs, and amplifies signals that are NOT common to both inputs. If the common-mode signals are not equal at the two differential inputs, they will appear at the differential amplifier output, distorting the desired brain-wave signal. The common-mode input impedance is denoted Z1,2. The differential input impedances are denoted Z1,3 and Z2,3.

Brain-wave biopotentials are very small. The impedance of the human body is very large, and the external interference voltages (hum and noise) are very large. Measuring brain-wave biopotentials is extremely difficult because the biopotentials on the scalp are 100 microvolts or less, and the external interference voltages can be 10 volts or more. The impedance of the human head is about 1000 ohms to 10,000 ohms. The high impedance of the head requires a differential amplifier with a very high input impedance (typically 100,000 ohms to 100 million ohms), so that the biopotential voltage source is not loaded. The amplifier also must NOT contribute any spurious signals to the biopotential signals. An instrumentation amplifier is required to satisfy these requirements.

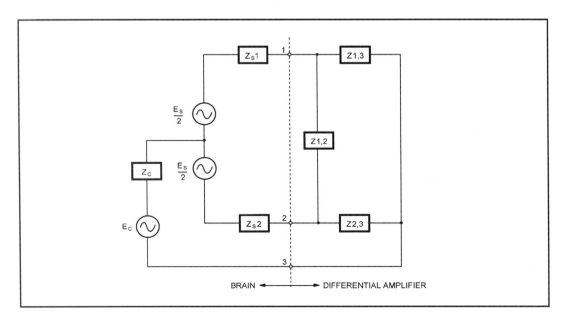

Figure 13-1. Electrical model of the brain.

Alpha Brain-Wave Feedback Monitor

You need an instrumentation amplifier to accurately measure brain-wave biopotentials. An instrumentation amplifier is a high-gain, low-noise, high-input impedance differential amplifier. The different brain waves are very close in their frequency ranges. A switchable, four-pole bandpass filter is used to tune the center frequency of the theta, alpha and beta bands. The bandpass filter makes the selection of a particular brain wave much easier.

The skin surface brain voltages are coupled via three probes to the input of the brain-wave monitor. The front end is a high-input impedance, high-gain, low-noise instrumentation amplifier. The amplified brain-wave frequencies are selected by a four-pole bandpass filter. The brain-wave monitor is battery operated to eliminate shock hazards to the patient or experimenter. For the same reason, optoisolators must be used if the brain waves are to be displayed on an oscilloscope.

The Brain as a Voltage Source

Different brain waves occur depending on the mental state of a person. *Table 13-1* lists different brain waves and their characteristics.

Biopotentials are present in all living organisms. In human beings, these small voltages are generated by the activity of nervous system sensors, muscles and nerves.

Detection of brain-wave biopotentials is complicated by the minute signal voltages, high level of external interference in the form of noise and hum, and the high impedance of the human body. The voltages being measured are of the order of 10 to 100 microvolts, while the external noise may reach 10 volts or more.

The large external fields can be screened out by a differential amplifier. The same amplifier can amplify the small biopotentials. The differential amplifier rejects signals that are common to both inputs while amplifying difference signals between the two inputs.

The scalp has a typically high impedance of 1000 to 10,000 ohms. The differential amplifier must not load the signal source; therefore, it must have a high input impedance of at least 100,000 ohms. The amplifier must not add spurious signals to the biopotentials.

IC Design Projects

Biofeedback

Biofeedback is closed-loop system. When the required brain waves occur, a tone is generated, signaling the user that he or she is generating the desired brain waves. With practice, the user develops limited control over his or her brain by controlling the sound pattern of the tone generator. When a person meditates, he or she is generating alpha brain waves.

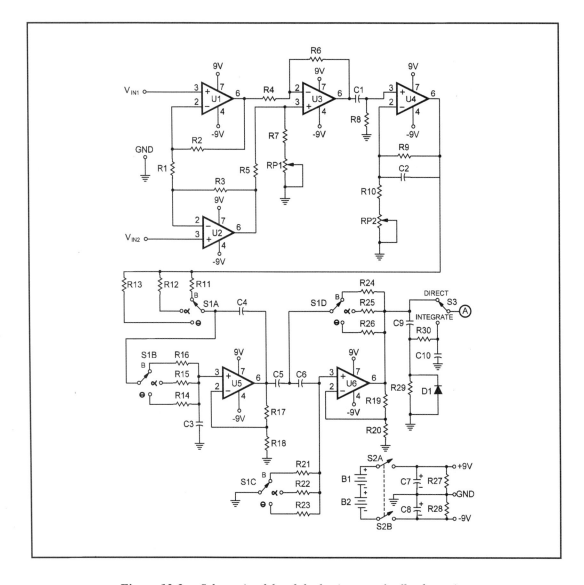

Figure 13-2a. Schematic of the alpha brain-wave feedback monitor.

Alpha Brain-Wave Feedback Monitor

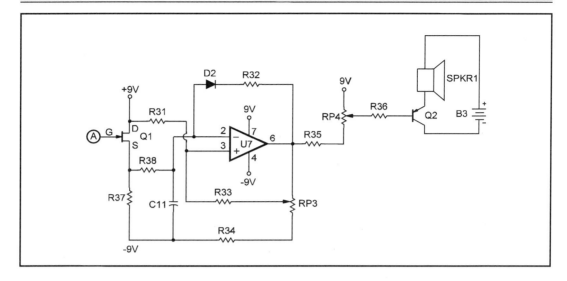

Figure 13-2b. Schematic of the alpha brain-wave feedback monitor (continued).

Circuit Operation

The schematic of the alpha brain-wave feedback monitor is shown in *Figure 13-2*. The front end is an instrumentation amplifier. It consists of U1, U2, U3 and U4, and associated components. The fourth-order bandpass filter is made up of U5, U6 and associated components. The tone generator is an AM/FM modulator consisting of FET Q1, operational amplifier U7 and associated components. Transistor Q2 is configured as an audio amplifier. In the direct position, the instantaneous waveform modulates an adjustable but continuous tone. In the integrate position, only the peaks of the filtered waveform trigger the tone. The threshold control RP3 can be adjusted so that the tone is only present when the desired signal is present. RP4 is a volume control.

U1 and U2 are FET operational amplifiers. This type of op-amp is used to provide a very high input impedance and excellent noise rejection. The differential amplifier amplifies the difference signal while providing unity gain for the common-mode signal. Each half of the differential amplifier (that is, U1 and U2) provides a gain of 47 to the difference input signal.

The null stage consists of U3. When RP1 is adjusted properly, the sum of resistances RP1 and R7 equals each of the resistances R4, R5 and R6, and the common-mode signal or noise eliminated. The difference input signal is passed through with unity gain. The component tolerances are critical for this stage.

IC Design Projects

The variable gain stage consists of U4. It also provides a low frequency and a high frequency limit to the brain-wave monitor. The low-frequency limit is necessary so that stray electron movements do not cause a varying DC offset level to the amplified brain waves. The high-frequency limit is set at a higher level than that required to measure brain waves. This provides a more stable circuit because the op-amps are compensated below their cutoff frequency. The gain of this stage can be varied by gain control RP2, from 4.5 to 100. The high-pass filter consists of C1 and R8. The cutoff frequency is 1.6 Hz. The low-pass filter consists of C2 and R9. The cutoff frequency is 34 Hz. U4 is an amplifier at frequencies below 343 Hz, and it is an integrator at frequencies above 34 Hz.

The bandpass filter consists of U5, U6 and associated components. The bandpass filter is actually a Sallen and Key low-pass filter in cascade with a Sallen and Key high-pass filter. The bandpass filter provides a gain of about 2.6. The fourth-order bandpass filter has roll-off slopes of 12dB per octave. The brain-wave frequency is selected by S1, which is a four-pole, three-position switch.

Switch S3 is used to select the direct or integrate mode. Diode D1 is a shunt rectifier, and

---

All resistors are 1/4W @ 5% unless otherwise noted.
All capacitors are rated 16 volts.

R1, R36:	1000 ohms
R2, R3:	47 kohms
R4, R5, R6:	3900 ohms
R7:	3300 ohms
R8, R33:	100 kohms
R9, R31:	470 kohms
R10:	4700 ohms
R11, R16, R21, R24:	18 kohms
R12, R15, R22, R25:	33 kohms
R13, R14, R23, R26:	68 kohms
R17, R19:	1800 ohms
R18, R20:	3000 ohms
R27, R28, R32:	22 kohms
R29:	1.0M ohms
R30:	4.7M ohms
R34, R35:	10 kohms
R37:	39 kohms
R38:	1.5M ohms
RP1:	1000 ohms potentiometer
RP2:	100 kohms potentiometer
RP3:	50 kohms potentiometer
RP4:	10 kohms potentiometer
C1:	1 uF
C2:	0.01 uF
C3, C4, C5, C6:	0.47 uF
C7, C8:	100 uF
C9, C10:	0.1 uF
C11:	0.001 uF
U1, U2:	N5556
U3-U7:	LM741
D1, D2:	1N4004
Q1:	TIS58
Q2:	2N4250
B1, B2, B3:	9-volt battery
SPKR1:	8 ohms
S1:	4P3T
S2:	DPDT
S3:	SPDT

Table 13-2. Parts list for the alpha brain-wave feedback monitor.

capacitor C10 is a low-pass filter. The signal is passed to FET Q1, which operates as a unity gain source follower. Operational amplifier U7 is configured as a multivibrator that is normally saturated when the output signal is near the positive supply voltage. When capacitor C11 charges through resistor R38 to a voltage greater than the level provided by the voltage divider (consisting of resistors R31, R33, R34 and RP3), integrated circuit U7 saturates due to positive feedback. Capacitor C11 discharges through diode D2 until U7 flips back to its previous state. The signal from Q1 varies the charge on C11, and thus modulates the tone.

Transistor Q2 is a course follower, which provides a low impedance to drive the speaker without overloading the multivibrator. A separate 9-volt battery, B3, is used for the speaker to avoid feedback.

The main power supply consists of two 9-volt batteries, B1 and B2. As the batteries age, their voltages may not be equivalent. Components C7, C8, R27 and R28 form a voltage divider which ensures that the circuit ground is at "half potential." The power switch, S2, is a two-pole, two-position switch.

Construction

The alpha brain-wave feedback monitor may be built on a piece of perfboard. *Table 13-2* is the parts list for the alpha brain-wave feedback monitor. Install the ICs and the electrolytic capacitors with the proper orientation. You should use IC sockets. IC sockets prevent damage to the ICs caused by soldering iron heat. It is also easier to replace ICs when IC sockets are used. The brain-wave monitor may be wire-wrapped. Try to keep the wire lengths as short as possible.

Electrodes

A transducer is any device that converts energy from one form to another form. The loudspeaker and the thermocouple are two well-known transducers. A transducer is required to take surface skin voltages and feed them to the input of the alpha brain-wave feedback monitor. These probes are not strictly transducers, because they do not change one form of energy to another form of energy. They are more accurately called "conditioners." The measurement of these skin potentials has become an important tool to investigators of psychophysiological phenomena. To get a meaningful recording of these responses, stable transducers capable of "conditioning" the signals are required.

Brain-wave signals originate in the brain, pass through cerebral fluids, and reach the surface electrodes. Electrode cream is smeared on the electrodes to power the contact resistances between the scalp and the electrodes. The biopotential signals are then fed into the instrumentation amplifier inputs.

Electrodes couple the small biopotentials to the input of the instrumentation amplifier. The electrodes should not generate short-term voltages (tiny noise spikes) or long-term voltages (offset or drift).

There are several types of skin surface electrodes. Some are disposable; use them once then throw them away. Semidisposable electrodes can be used a few times. The thin silver plating is lightly sanded after each use. When there is no silver plating left on them, the electrodes are thrown away.

There are also permanent electrodes. They are usually the silver-silver chloride or pellet type. They are carefully compounded to keep offset voltage and polarization low. The permanent electrode has a larger sensing surface, which translates to a low source impedance and lower noise. The recessed element also gives the electrodes a lower susceptibility to motion artifacts. The silver-silver chloride pellet electrode also provides extremely stable, precise and dependable measurement that is virtually drift free.

Balancing the Instrumentation Amplifier

Potentiometer RP1 is used to trim the gain of the noninverting input of the differential amplifier (U3 and associated components) so that its gain matches that of the inverting input of the differential amplifier. When the gains are equal, the common-mode rejection ratio is maximum. Install new batteries and turn the brain-wave monitor on. Adjust potentiometer RP1 for a maximum common-mode rejection ratio.

A 3-volt to 4-volt common-mode signal is fed into both inputs. The two inputs are tied together. A 10,000-ohm resistor is placed between the common-mode signal and ground potential. Adjust RP1 for the smallest signal on the output of U4, as measured with either an oscilloscope or an AC voltmeter.

If a signal generator is not available, connect the electrodes through a 10,000-ohm resistor to ground potential. Touch the leads. You should see 60 Hertz of noise on the oscilloscope. Adjust RP1 for a minimum noise display on the oscilloscope.

Using the Alpha Brain-Wave Feedback Monitor

The placement of the electrodes on the head is critical. For the bandwidth settings on the alpha brain-wave feedback monitor, different electrode placements yield different brain waves.

When placing the electrodes on the head, it's necessary to use a lot of electrode cream. This ensures a good electrical contact between the skin and the plated surface of the electrode.

The heartbeat may be monitored by holding one electrode in each hand and clipping the ground electrode to an earlobe. The bandpass filter switch should be set to the theta or lowest frequency range.

The output of the brain-wave monitor may be fed to the input of an oscilloscope or to the input of a chart recorder. An optoisolator device should be used between the brain-wave monitor and any AC-operated equipment to prevent shock hazards.

Place one electrode above one eyebrow. Place a second electrode in line with the first one at the back of the head. The third electrode is connected to an earlobe. The third electrode is the ground electrode. A band may be tied around the head to hold the electrodes in place. Place the mode switch, S3, in the direct position and place the bandpass switch, S1, in the alpha position. Adjust RP3 and RP4 for a pleasing tone and volume. A beep should be heard when you blink your eyes. Once you are able to produce alpha brain waves, set switch S3 in the integrate position. Adjust RP3 for no tone when the eyes are opened. Close your eyes and try to generate a tone by generating alpha brain waves.

When you are finished using the brain-wave monitor, wipe off the electrode cream from the electrodes. If stainless steel electrodes are used, sand them lightly and clean them with alcohol.

The brain-wave monitor should provide years of reliable service.

Possible Improvements

The alpha brain-wave feedback monitor, as it stands, is a good circuit. However, there are a few possible improvements.

A notch filter, designed to eliminate 60 Hz noise, can be installed between the null stage, U3, and the variable gain stage, U4. This will reduce AC hum as well.

The monitor circuit generates noise. This is due to the noisy LM741 op-amps used. U3-U6 should be replaced with low noise and uncompensated op-amps. The new op-amps should be externally compensated to a cutoff frequency of about 100 Hz. This would reduce circuit-generated noise as well as increase the overall circuit stability. You should verify the pin configuration of the new op-amps and to wire them properly into the circuit.

The active bandpass filter is a fourth-order filter. To increase the roll-off slope, another bandpass filter can be cascaded with the original bandpass filter. If the overall bandpass filter gain is too high, the circuit will saturate. The gain of the original bandpass filter can be reduced by half, then two identical filters will restore the gain to its original value.

An optoisolator device may be incorporated into the circuit at the required output to be displayed on an oscilloscope. This will protect the subject from shock hazards.

Chapter 14
◆ IC Stereo Preamplifier ◆

This stereo preamplifier was designed using op-amps. The preamplifier features an RIAA preamplifier to amplify the small signals of a magnetic phono cartridge or CD player. The stereo preamplifer also features bass and treble boost, and cut-tone control circuits. The sound may be tailored to suit the listening room acoustics by adjusting the bass and treble controls.

The preamplifier incorporates switching functions that allow recording with a tape or cassette deck, as well as with a VCR. CD-quality recordings can be made with a tape deck. A DAT recorder is not required for CD-quality recordings with this stereo preamplifier.

Each channel has its own volume control. The music balance is adjusted with the left and right channel volume controls.

Circuit Description

The schematic of one channel of the preamplifier is shown in *Figure 14-1*. Transformer T1 steps down the household AC line voltage to a lower AC voltage. Diodes D1 - D4 rectify the AC voltage into a pulsating DC voltage. Capacitors C9, C10, C13 and C14 smooth the pulsating DC voltage into a low-ripple, fluctuating DC voltage. Integrated circuits U4 and U5 are positive and negative voltage regulators, respectively. Capacitors C12 and C16 improve the transient responses of the IC voltage regulators.

The preamplifer consists of an RIAA preamplifier, a bass and treble tone control circuit, and a variable gain stage.

If you are using a magnetic phono cartridge, the output is usually less than 10 millivolts. The bass cut is used during the recording process to prevent excessive groove modulation at low frequencies (obviously, with CDs, this is not a problem). The RIAA preamplifier must provide RIAA equalization, and it must also provide high gain at low frequencies.

IC Design Projects

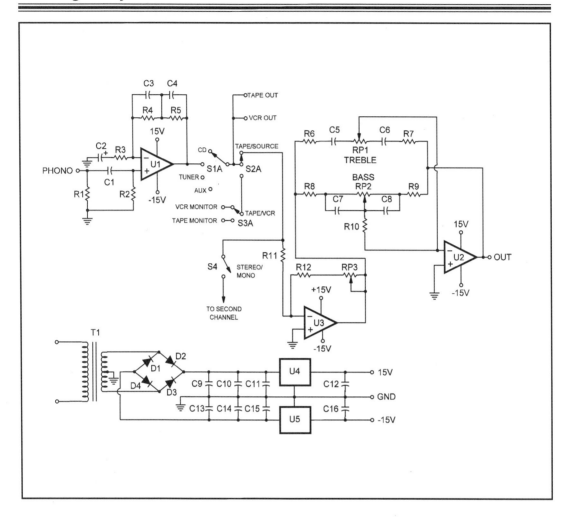

Figure 14-1. Schematic of the IC stereo preamplifier (one channel shown).

The RIAA preamplifier stage consists of integrated circuit U1, resistors R1 - R5, and capacitors C1 - C4. The op-amp is configured as a noninverting amplifier. Resistor R1 loads the magnetic cartridge. The voltage generated by the magnetic cartridge is coupled to the noninverting input of the op-amp by capacitor C1. Resistor R2 balances the output offset, due to the bias current drawn through feedback resistors R4 and R5, by the inverting input of the op-amp. Capacitor C2 decreases the gain of the circuit at very low frequencies. This is necessary to eliminate turntable rumble when using a phono cartridge (but unnecessary for a CD player). Capacitor C2 also decreases the output offset due to the input offset voltage and offset current. The proper RIAA equalization is established by the two RC (R4, C3 and R5, C4) networks in the negative feedback loop.

Page 138

Switch S1 is used to select the signal source. Switch S2 is used to select either the tape source or the signal source (phono or CD). Switch S3 is used to select either the tape recorder or the VCR. Switch S4 connects the two channel inputs together when the signal source is a monophonic signal source.

The variable gain stage consists of IC U3 and resistors R11, R12 and RP3. The op-amp is configured as an inverting amplifier. Potentiometer RP3 is used to adjust the gain. The gain is variable from unity to 11.

The tone-control stage consists of IC U2, resistors R6 - R10, potentiometers RP1 and RP2, and capacitors C5 - C8. Potentiometer RP1 is used to adjust the treble boost or the treble cut. Potentiometer RP2 is used to adjust the bass boost or the bass cut. The op-amp is configured as an inverting amplifier. The treble tone control circuit is a high-pass filter whose gain (or attenuation) is varied with potentiometer RP1. The cutoff frequency is about 5 kilohertz. The bass tone control circuit is a low-pass filter whose gain (or attenuation) is varied with potentiometer RP2. The cutoff frequency is about 50 Hertz.

The bass control circuit can boost or cut frequencies less than 30 Hertz by +/-6 decibels. The tone control circuit can boost or cut frequencies above eight kilohertz by +/-12 decibels. The frequency response for the phono input is 10 Hertz to 100 kilohertz. The frequency response for the other (CD) inputs is DC (or zero Hertz) to 100 kilohertz. The input and output signals are in phase because the RIAA preamplifier is a noninverting stage. The tone control and variable gain circuits are inverting stages. The result is the input and output signals are in phase. Phase distortion is therefore kept to a minimum.

Construction

The stereo preamplifier is easy to build, as can be seen from the parts list in *Table 14-1*. Although only one channel is shown in the schematic of the preamplifier in *Figure 14-1*, all part numbers above 100 refer to the second channel part numbers.

The preamplifier can be built on a piece of perfboard. Take care to orient all diodes, capacitors and ICs properly. IC sockets should be used.

IC Design Projects

The IC stereo preamplifier should work the first time. Take care when soldering to avoid cold solder joints and short circuits created by solder bridges. Most problems are caused by cold solder joints and solder bridges.

All resistors are 1/4W @ 5% unless otherwise noted.	
All capacitors are rated 25 volts.	
All part numbers above 100 refer to the second channel parts.	
R1, R4, R101, R104:	47k
R2, R102:	510k
R3, R103:	470 ohm
R5, R105:	470k
R6, R7, R106, R107:	1.5k
R8, R9, R108, R109:	12k
R10, R110:	4.7k
R11, R12, R111, R112:	10k
RP1, RP2:	100k linear dual potentiometer
RP3, RP103:	100k audio taper potentiometer
C1, C101:	0.1 uF
C2, C102:	47 uF
C3, C103:	0.0015 uF
C4, C104:	680 pF
C5, C6, C105, C106:	4.7 nF
C7, C8, C107, C108:	47 nF
C9, C10, C13, C14:	1000 uF
C11, C15:	0.33 uF
C12, C16:	1.0 uF tantalum
D1-D4:	1N4004
U1, U101:	MC1456
U2, U3, U102, U103:	MC741
U4:	LM7815
U5:	LM7915
T1:	30 V.C.T., 1 amp secondary
S1:	2P4T rotary switch
S2, S3:	2P2T
S4:	SPST

Table 14-1. Parts list for the IC stereo preamplifier.

Part Four
♦ Phase-Locked ♦
♦ Loop ♦

Chapter 15
◆ Phase-Locked Loop ◆

Phase-locked loop (or PLL) systems are designed to synchronize one sinusoidal waveform to another sinusoidal waveform. PLL circuits are used in everything from color television, to missile tracking and space telemetry.

A basic PLL circuit consists of a voltage-controlled oscillator (or VCO), a phase comparator, and a low-pass loop filter, as shown in *Figure 15-1*. The VCO generates a reference signal. The phase comparator compares the frequencies of the reference signal and the input signal. The phase comparator generates an error voltage, which is proportional to the difference between the reference signal frequency and the input signal frequency. The low-pass loop filter smooths the pulsating error voltage into a DC error voltage, which is used to control the oscillation frequency of the VCO.

The error voltage causes the VCO to change its frequency of oscillation until it matches the frequency of the input signal. This is known as the capture process. The PLL is locked or phase-locked to the input signal.

The PLL automatically tracks changes in the input frequency that fall within a range of frequencies known as the lock range. The capture range is the band of frequencies over which the PLL can hunt and capture an incoming signal. The lock range is always larger than the capture range.

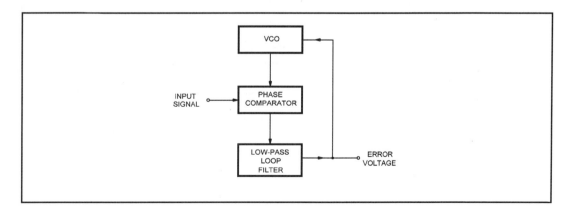

Figure 15-1. Block diagram of a basic PPL.

The low-pass loop filter limits the speed with which the PLL can track changes in the input frequency. It also limits the capture range. The low-pass loop filter prevents high frequency noise spikes from adversely affecting the operation of the PLL. The charge stored on the loop filter's capacitor speeds up the recapture of a signal that is temporarily lost due to a noise spike or other transient.

The bandwidth of the PLL is controlled by the bandwidth of the low-pass loop filter. It can be made very small to reject high frequency noise spikes. If the bandwidth of the PLL is too small, it will never capture and lock onto the input signal. The PLL can lock onto the input signal if its frequency is within one bandwidth (of the loop filter) of the VCO's frequency. The PLL operates nonlinearly when it is not locked onto the input signal.

The PLL is used in many types of communication and measurement equipment. There are several PLL ICs available to the circuit designer.

PLL Commercial Applications

A PLL circuit can be used to design a stereo demodulator as shown in *Figure 15-2*. There is a divide-by-N circuit following the VCO in the feedback loop. The VCO frequency is therefore exactly N times the input signal frequency. This technique can be used in designing frequency synthesizers.

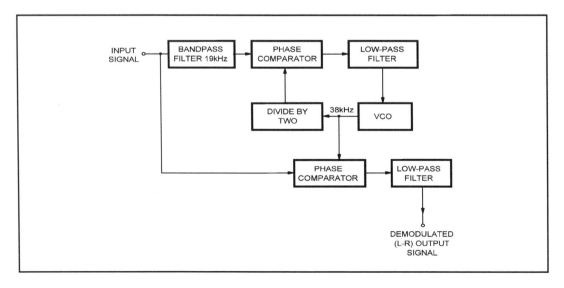

Figure 15-2. Block diagram of a PPL stereo demodulator.

Phase-Locked Loop

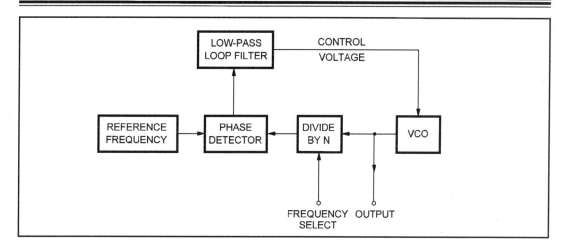

Figure 15-3. Block diagram of a PLL frequency synthesizer.

The PLL permits the digital control of a frequency synthesizer in linear frequency steps. *Figure 15-3* is a block diagram of a frequency synthesizer. The PLL frequency accuracy can be very high if a crystal oscillator reference is used. The PLL synthesizer divides the signal frequency by an integer, N, and compares the result to a stable and precise reference frequency in a phase detector. The phase detector output voltage is the error voltage, fed back to the VCO through a loop filter. The loop maintains the output of the programmable divider at the same frequency as the reference signal. The output frequency, F, is F = Nf, where f is the reference frequency.

Phase comparison can take place at high speeds. Frequency corrections can also be made at high speeds. A control voltage correction can only be made once every cycle of the reference frequency. The correction at the output of the phase detector is therefore a sampled function with a frequency equal to the reference frequency. The phase detector output or error signal is filtered then used to modulate the VCO. These modulations cause sidebands to be generated.

There are several PLL ICs available to the circuit designer.

LM565 PLL Integrated Circuit

The LM565 is a general-purpose PLL integrated circuit. A block diagram of the LM565 PLL integrated circuit is shown in *Figure 15-4*. It consists of a stable and linear VCO, as well as a double-balanced phase detector.

IC Design Projects

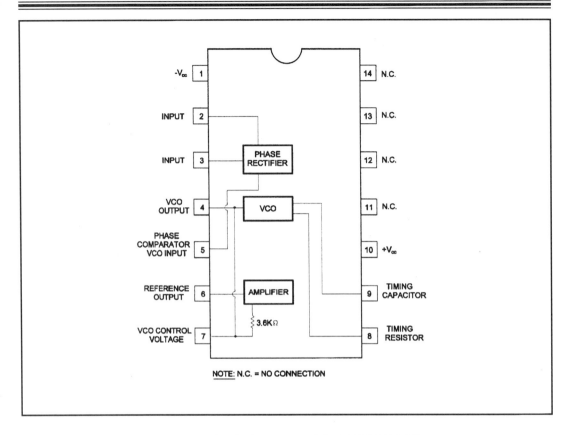

Figure 15-4. Block diagram of the LM565 PLL IC.

The VCO frequency is set with an external resistor and an external capacitor. The bandwidth, response speed, capture range and pull-in range of the closed loop system can be adjusted by an external resistor and an external capacitor. A digital frequency divider may be inserted between the VCO and the phase detector to obtain frequency multiplication.

Pin 1 is connected to the negative rail of a dual or bipolar power supply. The negative rail must be at least -6 volts and not more than -12 volts.

Pins 2 and 3 are the input terminals to the phase detector. The input signal is usually connected to pin 2. Pin 3 is usually connected to ground.

Pin 4 is the output terminal of the VCO. The frequency of oscillation, F, is set by external resistor R1 and external capacitor C1, shown in *Figure 15-4*: $F = 0.3/R1C1$

Phase-Locked Loop

Pin 5 is the phase comparator VCO input. Pins 4 and 5 are usually connected together. If frequency multiplication is required, pin 4 is connected to the input of a digital frequency divider, and the output of the digital frequency divider is connected to pin 5.

Pin 6 is the reference output terminal. It is a DC reference voltage that is close to the DC potential of pin 7.

Pin 7 is the demodulated output terminal. If a resistor is placed between pins 6 and 7, the gain of the amplifier (see *Figure 15-4*) is reduced with little change in the DC voltage level of the output voltage. The lock range can therefore be decreased.

An external resistor is connected between pin 8 and the positive rail of the dual power supply. An external capacitor is connected between pin 9 and the negative rail of the dual power supply.

Pin 10 is connected to the positive rail of a dual or bipolar power supply. The positive rail must be at least +6 volts and not more than +12 volts.

There are no connections required to pins 11, 12, 13 and 14.

A low-pass loop filter consists of the internal 3.6-kilohm resistor and external capacitor C2, which is connected between pins 7 and 10, as shown in *Figure 15-4*. The low-pass loop filter determines the capture characteristics of the loop. The VCO cannot operate separately because its control voltage input terminal is connected to the amplifier output through the internal, 3.6-kilohm, low-pass loop filter resistor.

The LM565 can be powered by a single-ended power supply of at least +12 volts, and not more than +24 volts. The LM565 PLL integrated circuit can phase-lock to any input signal that is within the range of +/-60 percent of the oscillation frequency of the VCO.

Applications for the LM565

The LM565 PLL can be used in many applications, such as frequency-shift keying (FSK), telemetry receivers, tone decoders and FM discriminators.

An FM detector circuit is shown in *Figure 15-5*. No transformers or tuned circuits are required. The input signal voltage must be at least one millivolt. The PLL bandwidth

IC Design Projects

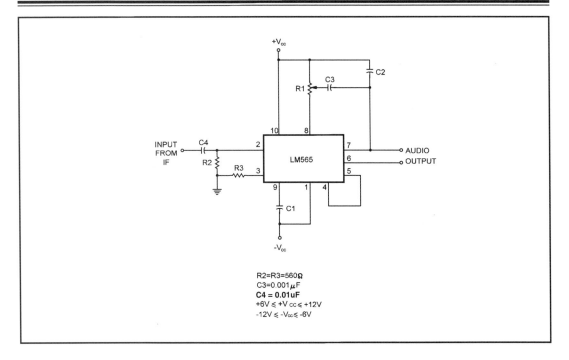

Figure 15-5. Schematic of an FM demodulator.

is usually between 2 percent and 10 percent of the intermediate frequency (or i-f) for FM detection. Components R1 and C1 set the VCO to near the desired frequency. Capacitor C2 is the loop-filter capacitor which determines the capture range. The capture range is the range of frequencies over which the loop can lock on an input signal, which is initially out of lock. The LM565 has an upper frequency limit of 500 kilohertz.

CMOS PLL Integrated Circuit

The CD4046 CMOS PLL integrated circuit consists of a VCO, two phase comparators, and an amplifier, as shown in *Figure 15-6*. There is also a 5.2-volt zener diode to provide power supply regulation, if required.

Pin 1 is an output of phase comparator II. It is high when phase comparator II is locked onto the input signal. Phase comparator II consists of four edge-triggered flip-flops, control gates, and a three-state output stage. This phase comparator is sensitive to noise but it does not readily lock onto the harmonics of the VCO frequency of oscillation. It can accept input pulses of any duty cycle.

Phase-Locked Loop

Pin 2 is the output of phase comparator I. This phase comparator is an EXCLUSIVE-OR gate that is immune to noise. Phase comparator I has a tendency to lock onto input signals whose frequencies are close to the harmonics of the VCO frequency of oscillation. This phase comparator requires a square wave input signal with a 50 percent duty cycle.

Pin 3 is an input terminal for both phase comparators.

Pin 4 is the output of the VCO. It is a symmetric square wave output.

Pin 5 is an inhibit input terminal. A low on pin 5 enables the VCO and the source follower. A high on pin 5 disables the VCO and the source follower to reduce power consumption.

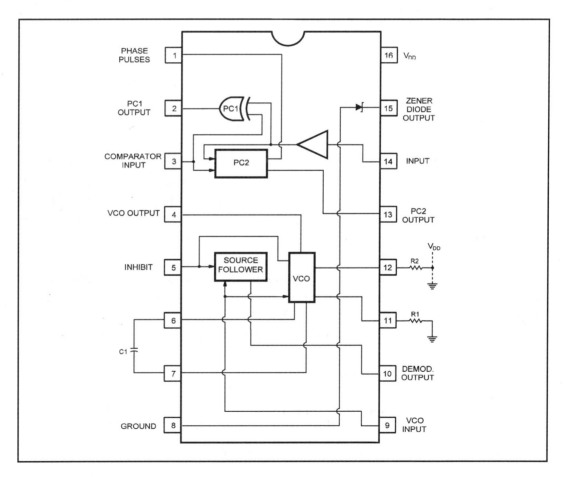

Figure 15-6. Block diagram of the CD4046 PLL IC.

Page 149

IC Design Projects

A capacitor (C1 in *Figure 15-6*) is connected between pin 6 and pin 7. It must be at least 50 pF, and determines the center frequency of the VCO.

Pin 8 must be connected to the ground potential of the circuit in which the CD4046 is used.

Pin 9 is the control voltage input terminal for the VCO. The VCO can oscillate at any frequency up to about one megahertz. The control voltage input terminal has a high input impedance, and it must be driven by a high impedance source.

Pin 10 is the demodulated output of the source follower. The high impedance source driving pin 9 may be monitored at pin 10 without loading the high impedance source. If pin 10 is used to drive an external circuit, a resistor of at least 10 kilohms must be connected between pin 10 and ground; otherwise, pin 10 should be left floating.

Pin 11 is connected to ground through resistor R1, as shown in *Figure 15-6*. It determines the center frequency of the VCO.

Pin 12 is connected through resistor R2 to any voltage between ground and the positive power supply rail. It also determines the center frequency of the VCO.

When only R1 is used, the VCO frequency can range from DC to frequency F1, where $F1 = 1/(R1(C1 + 32 \text{ pF}))$. The VCO operates at DC when the control voltage at pin 9 is connected to ground. It operates at F when the control voltage at pin 9 is connected to the positive power supply rail. Resistor R1 can have any value between 10 kilohms and 10 megohms.

Resistor R2 is optional, used when the minimum VCO frequency must exceed DC (or zero Hertz). When R2 is used, and pin 9 is connected to ground, the minimum frequency, F2, is $F2 = 1/(R2(C1 + 32 \text{ pF}))$. When R2 is used, and pin 9 is connected to the positive power supply rail, the maximum VCO frequency, F, is $F = F1 + F2$.

Pin 13 is the output of phase comparator II.

Pin 14 is the input to the input signal amplifier.

Pin 15 is a zener diode-regulated, 5.2-volt voltage source.

Phase-Locked Loop

Pin 16 must be connected to a positive power supply in the range of 3 volts to 18 volts.

Power consumption depends upon the VCO frequency and the percentage of time that the VCO is enabled. The CD4046 is suited for battery-operated circuits.

Problems

Problem 15-1. What does PLL stand for?
Problem 15-2. What is PLL used for?
Problem 15-3. Name the components of a PLL system and list their functions.
Problem 15-4. What is the capture process?
Problem 15-5. What is the lock range?
Problem 15-6. When is the PLL not operating linearly?
Problem 15-7. If a divide-by-ten circuit is used in a PLL loop with a reference frequency of 10 MHz, what is the output frequency?
Problem 15-8. In an LM565 circuit, R1 = 100 kilohms and C1 = 100 pF. What is the VCO frequency?
Problem 15-9. Why can't the VCO of an LM565 PLL operate separately?
Problem 15-10. What is the advantage to using an LM565 PLL integrated circuit in an FM detector circuit?
Problem 15-11. Define capture range.
Problem 15-12. What are the differences between the two phase comparators of a CD4046 PLL integrated circuit?
Problem 15-13. Calculate F1 if R1 = 100 kilohms and C1 = 68 pF in a CD4046 circuit.
Problem 15-14. Calculate F2 if R2 = 10 kilohms and C1 = 68 pF in a CD4046 circuit.
Problem 15-15. Calculate F if R1 = 100 kilohms, R2 = 10 kilohms and C1 = 68 pF in a CD4046 circuit.

Chapter 16
◆ Function Generator ◆

A function generator produces voltage waveforms, which can be used to test electronic circuits. The most popular waveforms are sine waves, square waves and triangular waves.

This function generator produces square waves and triangular waves. The square waves can be used to test digital circuits, and the triangular waves can be used to test analog circuits.

The function generator has a frequency control and a range switch. It can produce waveforms from about 4 Hertz to about 65 kilohertz. The amplitude of the waveform can be adjusted with the gain control. The function generator should provide years of reliable service.

Circuit Operation

The function generator circuit is shown in *Figure 16-1*. PLL integrated circuit U1 is the heart of the function generator. The frequency of operation of the VCO is determined by resistor R1, potentiometer R2 and capacitors C1 - C4. Potentiometer R2 is a frequency adjust control, and switch S1 is used to select the appropriate range capacitor.

With capacitor C1 switched in, the frequency range is roughly 4 Hertz to 40 Hertz. With capacitor C2 switched in, the frequency range is roughly 20 Hertz to 300 Hertz. With capacitor C3 switched in, the frequency range is roughly 200 Hertz to 4000 Hertz. With capacitor C4 switched in, the frequency range is roughly 4000 Hertz to 65,000 Hertz.

The required waveform is selected by switch S2. The selected waveform is fed through gain control R3, to the noninverting input of the operational-amplifier, U2, which is configured as a buffer amplifier. The buffer amplifier has a very high input impedance and a very low output impedance. The buffer amplifier is required because the relatively high output impedances of U1 pin 4 and U1 pin 9 must not be loaded.

IC Design Projects

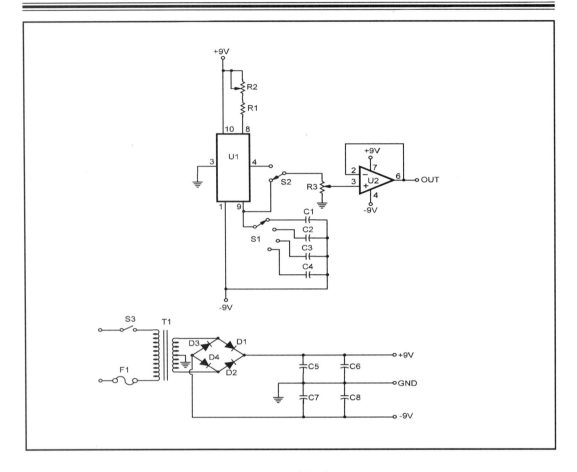

Figure 16-1. Schematic of the function generator.

The power supply is a bipolar or dual supply. Transformer T1 steps down the household AC line voltage to a lower AC voltage. Diodes D1 - D4 form a bridge rectifier. The bridge rectifier and the center-tapped secondary of the transformer are the heart of the bipolar or dual power supply. The pulsating DC voltages are smoothed by filter capacitors C5 - C8 into low-ripple fluctuating DC voltages.

Construction

The function generator is an easy project to build as can be seen from the parts list, which is given in *Table 16-1*. The function generator may be built on a piece of perfboard. Care should be taken to orient the ICs, diodes and capacitors properly. IC sockets are recommended. A defective IC can be replaced without desoldering if IC sockets are used.

Function Generator

```
All resistors are 1/4W @ 5% unless otherwise noted.
All capacitors are rated 16 volts

R1:     2.7k
R2:     50k potentiometer
R3:     100k potentiometer
U1:     LM565
U2:     LM1456 or LM741; see text
C1:     1.0 uF
C2:     0.1 uF
C3:     0.01 uF
C4:     470 pF
C5-C8:  1000 uF
D1-D4:  1N4004
T1:     12.6 V.C.T. secondary @ 1 ampere
S1:     SP4T, rotary switch
S2:     SPDT
S3:     SPST
F1:     1A slow blow
```

Table 16-1. Parts list for the function generator.

Care should be taken when soldering. Most problems are caused by cold solder joints and shorts created by solder bridges. If you are inexperienced with soldering, you should practice soldering on a scrap piece of perfboard. The function generator should work properly the first time it is powered up.

Testing and Use

The function generator can be built and tested in stages. The dual power supply should be built first. The power supply dual voltages should be about +9 volts and -9 volts. If not, verify that the diodes and filter capacitors are properly oriented. Check the transformer for continuity on its primary and secondary windings. Verify that F1 is not blown and that switch S3 is in its closed position.

When the bipolar power supply is working properly, build the PLL circuit, which consists of IC U1, resistor R1, potentiometer R2, capacitors C1 - C4, and switches S1 - S2. Verify that U1 pin 10 is connected to the positive power supply rail, that U1 pin 1 is connected to the negative power supply rail, and that U1 pin 3 is connected to ground. If an oscilloscope is available, verify the waveforms at U1 pin 4 and U1 pin 9. These waveforms are shown in *Figure 16-2*. There should be a waveform at all R2 and S1 settings. If an oscilloscope is not available, the waveforms may be verified with an analog DC voltmeter. Select capacitor C1 with switch S1, and adjust potentiometer R2 for the lowest possible frequency. The voltmeter needle should move back and forth, indicating that the waveform is present.

The buffer amplifier consisting of integrated circuit U2 and potentiometer R3 should be built next. There should be a square or triangular wave output depending on the position of switch S2. The amplitude of the output waveform can be adjusted with potentiometer R3. If there is a problem, verify that the op-amp is oriented and wired properly. Almost any op-amp may be used as long as the proper terminals of the op-amp are wired into the circuit. The function generator should provide years of reliable service.

The square wave is ideal for testing digital circuits. Potentiometer R3 is used to adjust the amplitude of the square wave to suit the digital circuit under test.

The triangular wave is suitable for testing analog circuits. Potentiometer R3 is used to adjust the amplitude of the triangular wave to suit the analog circuit under test.

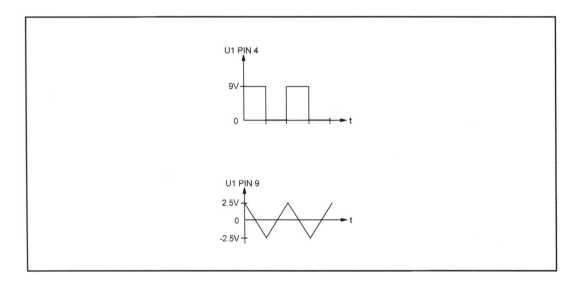

Figure 16-2. Waveforms of the function generator.

Chapter 17

◆ Siren ◆

The CD4046 is a very versatile PLL IC, and it can be used in the design of a wide variety of sound-effects generators. The VCO operating frequency can be swept over a wide range by varying the voltage applied to pin 9 of the CMOS PLL integrated circuit. Its output can be gated ON and OFF by applying a voltage to pin 5 of the CMOS PLL IC. This project produces a very realistic-sounding police siren.

The siren can be used as a stand-alone project or it can be used in conjunction with other circuits. The siren can be used as the alarm output of a home or automobile security system. It can also be used as a soundtrack for computer games.

This project may be battery powered because it draws very little current. The siren operates over a wide range of voltages, and can therefore be wired into the existing power supply of a computer, a security system, etc. Your imagination is the only limit to the uses in which this siren can be employed.

Circuit Operation

The siren uses three CMOS ICs, as shown in the schematic of *Figure 17-1*. Integrated circuit U1, resistor R1 and capacitor C1 are configured as an astable multivibrator.

Integrated circuit U2 is a quad bilateral switch. Only one of the four bilateral switches is used in the siren. Pin 1 and pin 2 are the switch input and output, respectively. Pin 13 is the switch control. When pin 13 is high, the switch is closed. When pin 13 is low, the switch is open.

The astable multivibrator (U1) controls the bilateral switch control terminal (U2 pin 13), which in turn controls the VCO input terminal (U3 pin 9). PLL IC U3 is configured as a VCO. When the NAND gate astable multivibrator's output is high, the bilateral switch is closed, connecting resistor R2 to the positive rail of the power supply. Capacitor C3 charges through resistor R2. When the NAND gate astable multivibrator's output is low, the bilateral switch is open. Capacitor C3 discharges through resistor R3.

IC Design Projects

The voltage drop across capacitor C3 is continuously changing. Therefore, the VCO frequency of oscillation is also continuously changing. The result is an "up-down" police siren effect.

The sound of the siren may be altered by experimenting with the component values. Components R1 and C1 control the cycle time of the siren. Components R4 and C2 control the frequency of the siren. A chirping sound is generated if a 0.001-microfarad capacitor is substituted for capacitor C2. Components R3 and C3 control the wail of the siren. A 1.0-microfarad capacitor may be substituted for capacitor C3.

Transistor Q1 is configured as an amplifier. Resistors R5 and R6 limit the current drawn through the transistor by the loudspeaker. The loudspeaker impedance must be at least 8 ohms. A high-power amplifier may be used. Components Q1, R5 and R6 are deleted, and U3 pin 4 is connected to the input of the power amplifier via a coupling capacitor.

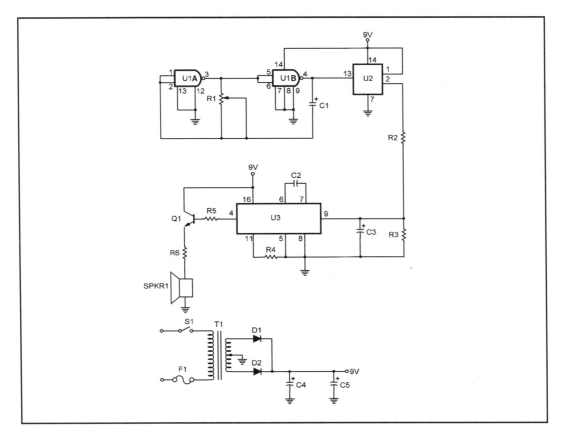

Figure 17-1. Schematic of the siren.

Transformer T1 steps down the household AC line voltage to a lower AC voltage. Diodes D1 and D2 rectify the AC voltage into a pulsating DC voltage. Capacitors C4 and C5 smooth the pulsating DC voltage into a low-ripple DC voltage.

Construction

The siren is a simple project, as can be seen from the parts list of *Table 17-1*. The project can be built on a piece of perfboard. IC sockets should be used because CMOS ICs are susceptible to damage due to static charges.

To prevent static damage to the CMOS ICs, a soldering iron with a grounded tip should be used. When installing the CMOS ICs, be sure to touch a ground with one hand while handling an IC with the other hand.

Care should be taken when soldering the components. Most problems are caused by cold solder joints and by short circuits created by solder bridges. If you are inexperienced with soldering, you should practice on a scrap piece of perfboard.

All resistors are 1/4W @ 5% unless otherwise noted.
All capacitors are rated at 16 volts.

R1:	100k potentiometer
R2:	330k
R3:	470k
R4:	100k
R5:	1k
R6:	220 ohms
C1:	10 uF tantalum
C2:	0.01 uF; see text
C3:	0.47 uF tantalum
C4, C5:	1000 uF
Q1:	2N3053
U1:	CD4011
U2:	CD4066
U3:	CD4046
D1, D2:	1N4004
T1:	12.6 V.C.T. secondary @ 1 ampere
S1:	SPST
F1:	1A slow blow
SPKR1:	8 ohms or greater

Table 17-1. Parts list for the siren.

If a chirping sound is preferred, a 0.001-microfarad capacitor may be substituted for capacitor C2. A 1.0-microfarad capacitor may be substituted for capacitor C3.

The project should work the first time. If not, check all connections for cold solder joints and short circuits created by solder bridges. All capacitors, ICs and diodes, and the transistor, must be properly oriented. The siren should provide years of reliable service.

Chapter 18
◆ Audible Logic Probe ◆

Most logic probes have visual displays. You must look at the visual display (usually LEDs) of the logic probe to determine the logic level of the test point of the circuit under test. Inadvertently, you can short two points of the circuit under test with the logic probe tip while looking at the visual display of the logic probe.

The audible logic probe gives you an audible indication of the logic level of the test point of the circuit under test. You no longer have to try to look at two places at once. You will no longer risk shorting two points of the circuit under test with the logic probe tip.

The audible logic probe uses one PLL CMOS integrated circuit. The circuit under test is not loaded by the logic probe.

Circuit Operation

The heart of the audible logic probe is the CD4046 PLL CMOS integrated circuit. The schematic of the audible logic probe is shown in *Figure 18-1*.

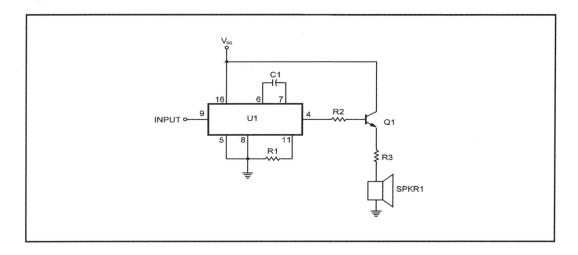

Figure 18-1. Schematic of the audible logic probe.

PAGE 161

IC Design Projects

Integrated circuit U1 is configured as a VCO. The frequency of operation of the VCO is determined by resistor R1 and by capacitor C1. The VCO input (pin 9) serves as the input of the logic probe. The voltage level of the input of the logic probe also determines the frequency of operation of the VCO.

The VCO output (pin 4) is connected to the base of transistor Q1 through a current-limiting resistor, R2. Transistor Q1 is configured as an amplifier. The output of the amplifier is fed to loudspeaker SPKR1 through a second current-limiting resistor, R3. The impedance of the loudspeaker must be 8 ohms or greater.

When the probe tip is connected to a low, there is no sound because the VCO is operating at DC (zero frequency). When the probe tip is connected to a high, the VCO operates at about 2500 Hertz; therefore, a 2500-Hertz tone is heard. When the probe tip is connected to an open circuit, a wailing tone centered around 1000 Hertz is heard. When the probe tip is connected to a pulse, a wailing tone centered around 800 Hertz is heard. The center frequency of the wailing tone depends upon the frequency of the pulse being measured.

The audible logic probe requires a power supply in the range of 3 to 18 volts. The audible logic probe is usually connected to the power supply of the circuit under test.

The schematic of the test circuit is shown in *Figure 18-2*. The LM555 timer IC, U1, is configured as an astable multivibrator. Using the parts listed in *Table 18-1*, the fre-

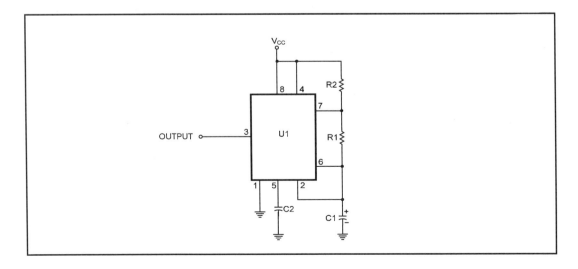

Figure 18-2. Schematic of the test circuit.

Audible Logic Probe

All resistors are 1/4W @ 5% unless otherwise noted.
All capacitors are rated at 25 volts.

R1:   100k
R2:   1k
C1:   0.1 uF
C2:   0.01 uF
U1:   LM555

Table 18-1. Parts list for the test circuit.

quency of operation is about 70 Hertz, and the duty cycle of the output waveform is about 50 percent. The test circuit component values are not critical.

Construction

The audible logic probe requires few parts. The parts list for the logic probe is listed in *Table 18-2*. It can be built on a piece of perfboard.

The transistor and IC must be properly oriented. You should use an IC socket, as it prevents solder heat damage to the PLL CMOS integrated circuit.

Care should be taken when soldering the circuit. Most problems are caused by cold solder joints and short circuits created by solder bridges. If you are inexperienced with soldering, you can practice on a scrap piece of perfboard beforehand.

Use care when working with CMOS ICs. A soldering iron with a grounded tip should be used. Touch one hand to ground while handling the CMOS IC with the other hand.

The audible logic probe can be housed in a speaker cabinet with the loudspeaker. Three lengths of wire should protrude through the speaker enclosure. The wires should be color-coded: red for Vcc, green (or black) for ground, and yellow (or white) for the audible logic probe input. The wires should be about two feet long, so that the circuit can be tested without moving the speaker enclosure. The red and green (or black) wires should have alligator clips connected to them, in order to facilitate connecting the wires to the power supply of the circuit under test. The yellow (or white) wire can be soldered to a sewing needle. A good solder connection is assured when the sewing needle plating is removed. The sewing needle serves as the audible logic probe tip. The audible logic probe should provide years of reliable service.

Testing and Use

The test circuit shown in *Figure 18-2* can be breadboarded to test the audible logic probe. The parts list for the test circuit is listed in *Table 18-1*.

The test circuit must be connected to a power supply in the range of 5 volts to 15 volts. The red wire of the audible logic probe is connected to the positive rail of the test circuit power supply. The green (or black) wire of the logic probe is connected to the ground of the test circuit.

When the audible logic probe tip touches pin 1 of the test circuit, there should be no sound, indicating a low. When the logic probe tip touches pin 4 (or pin 8) of the test circuit, a 2500-Hertz tone should be heard, indicating a high. When the logic probe tip touches pin 3, a wailing tone centered around 800 Hertz should be heard, indicating a pulse. The wailing tone center frequency depends upon the frequency of the pulse. If the audible logic probe tip touches an open circuit, a wailing tone centered around 1000 Hertz should be heard.

```
All resistors are 1/4W @ 5% unless otherwise noted.
All capacitors are rated at 25 volts.

R1:     100k
R2:     1k
R3:     220 ohms
C1:     0.01 uF
U1:     CD4046
Q1:     2N3053
SPKR1:  8 ohms or greater
```

Table 18-2. Parts list for the audible logic probe.

Part Five
♦ LM555 Timer IC ♦

Chapter 19
♦ The LM555 Timer IC ♦

The LM555 timer IC was introduced in 1972, and has been improved over the years. It is still one of the most popular ICs available.

The LM555 is a stable controller capable of producing accurate time delays, or oscillations. Terminals are provided for triggering or resetting. In the time delay mode of operation, the time is accurately controlled by two external components: one resistor and one capacitor. When the LM555 is configured as an oscillator in its astable mode, the free-running frequency and the duty cycle of the output waveform are accurately controlled with three external components: two resistors and one capacitor. The LM555 may be triggered and reset on trailing waveforms. The timer IC can source or sink 200 milliamperes, and can also drive TTL circuits.

The LM555 can time from microseconds to hours. It can operate in either the astable mode or the monostable mode. The duty cycle of the output waveform is adjustable. The timer IC is temperature stable, about 0.005 percent per degree Celsius. The output waveform is either normally on or off.

The LM555 timer IC is used in precision timing circuits, pulse-generating circuits, sequential timing circuits, time-delay generation circuits, missing pulse detector circuits, pulse-width modulation circuits, and pulse-position modulation circuits.

Inside the LM555

The LM555 is classified as a linear IC because it can be triggered by either a linear signal or a digital signal. The output of the LM555 is always a digital signal; that is, rectangular waves, square waves or pulses. The block diagram of the LM555 timer IC, 8-pin or mini-DIP, is shown in *Figure 19-1*. The mini-DIP package is the most popular LM555 timer package.

The LM555 consists of two comparators, a control flip-flop, and an output buffer stage.

IC Design Projects

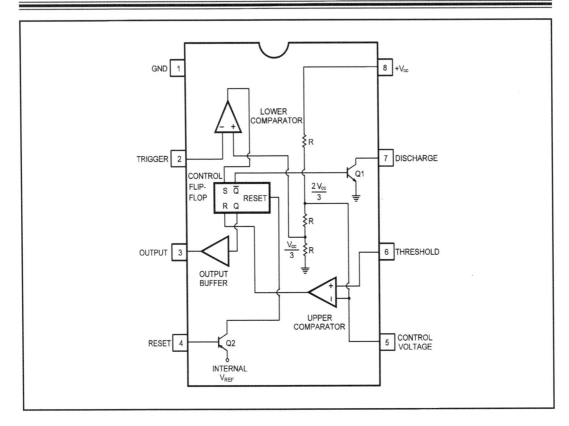

Figure 19-1. Block diagram of the LM555 timer IC.

The comparators are high-gain differential amplifiers. One input of each comparator is connected to a reference voltage. Three equal internal resistors provide the reference voltages. The other input of each comparator is connected to an input signal. These inputs are labeled THRESHOLD and TRIGGER. When the input signal of either comparator equals the reference voltage of the comparator in question, it changes its output state. The switching of the comparator causes the control flip-flop to change state as well.

The control flip-flop is a two-state binary circuit. The two states are SET and RESET. The upper comparator causes the control flip-flop to set, and the lower comparator causes the control flip-flop to reset. The control flip-flop generates the on-off pulses used at the output. The output of the control flip-flop is applied to an output buffer stage. The output buffer stage increases the current sourcing and current sinking capability of the LM555 timer IC.

The control flip-flop also drives the open-collector transistor, Q1. Transistor Q1 is connected to the external timing capacitor, and discharges the timing capacitor. Transistor Q2 accepts an input pulse that resets the control flip-flop. The reset signal is used to terminate the output pulse prior to the actual completion of the timing cycle, which is controlled by the external timing capacitor.

Pin 1 is connected to the ground of the circuit in which the LM555 timer IC is used.

Pin 2 is connected to the input signal of the lower comparator. Pin 2 is labeled TRIGGER.

Pin 3 is the output of the LM555 timer IC. It can sink or source 200 milliamperes of current.

Pin 4 is connected to the base terminal of the reset transistor, Q2. Pin 4 is labeled RESET.

Pin 5 is connected to the reference voltage terminal of the upper comparator.

Pin 6 is connected to the input terminal of the upper comparator. Pin 6 is labeled THRESHOLD.

Pin 7 is connected to the collector of the open-collector transistor, Q1. Q1 is used to discharge the timing capacitor.

Pin 8 is connected to the positive rail of the power supply. The power supply voltage must be in the range of 5 volts to 15 volts.

Astable Operation

An astable multivibrator is a free-running oscillator. It generates a continuous stream of rectangular on-off pulses that switch between the high and low states. The timing components determine the frequency of oscillation of the free-running oscillator.

The LM555 timer IC is configured as an astable multivibrator, as shown in *Figure 19-2*. The trigger and threshold inputs (pin 2 and pin 6) to the two comparators are connected together, and to the external timing resistor, R2, and capacitor, C1. Timing

IC Design Projects

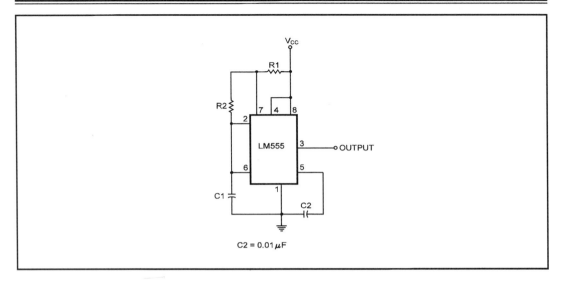

Figure 19-2. LM555 configured as an astable multivibrator.

capacitor C1 charges toward the supply voltage through external resistors R1 and R2. The discharge terminal (pin 7) is connected to the junction of resistors R1 and R2.

Timing capacitor C1 is initially discharged. The trigger and threshold inputs are initially zero volts. The lower comparator sets the control flip-flop. The flip-flop output switches high. Transistor Q1 turns off. Capacitor C1 begins to charge to 2/3 of the supply voltage, after which the upper comparator resets the control flip-flop. The flip-flop output switches low. Transistor Q1 turns on, effectively connecting resistor R2 to ground. Timing capacitor C1 discharges to 1/3 of the supply voltage through resistor R2, after which the lower comparator is triggered, again resetting the control flip-flop. The process is repeated, and the control flip-flop is repeatedly set and reset. The resulting waveforms are shown in *Figure 19-3*.

The charge time, T1, is T1 = 0.693(R1 + R2)C1.

The discharge time, T2, is T2 = 0.693 x R2 x C1.

The frequency of operation, F, is F = 1/(T1 + T2). F = 1.44/((R1 + 2R2)C1)

The duty cycle is the percentage of the total period in which the output pulse is on or high. The astable multivibrator can have a duty cycle from about 55 percent to about 95 percent, depending on the values of the external resistors, R1 and R2. The duty cycle, DC, is DC = T1/(T1 + T2) = (R1 + R2)/(R1 + 2R2)

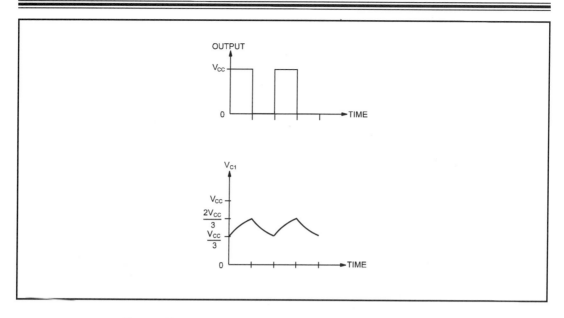

Figure 19-3. Waveforms for the LM555 astable multivibrator.

The maximum duty cycle occurs when resistor R1 is as small as possible. Resistor R1 also limits the discharge current available at pin 7 of the LM555 timer, to the 200-milliampere current rating of the internal discharge transistor. The minimum value of resistor R1 is R1 > Vcc/0.2, where Vcc is the power supply voltage.

Monostable Operation

A monostable multivibrator has only one stable state: the off state. Whenever the monostable is triggered by an input pulse, it switches to its unstable state for a time, which is determined by the timing components. The monostable circuit then returns to its stable state. The monostable circuit generates a single pulse of a fixed time duration each time it receives an input trigger pulse. The monostable multivibrator is also called a one-shot multivibrator.

The LM555 timer can be configured as a monostable multivibrator, as shown in *Figure 19-4*. The threshold input of the upper comparator is connected to the junction of the external timing components, R1 and C1. The internal discharge transistor, Q1, is also connected to the junction of the external timing components. The trigger input is applied to the input of the lower comparator.

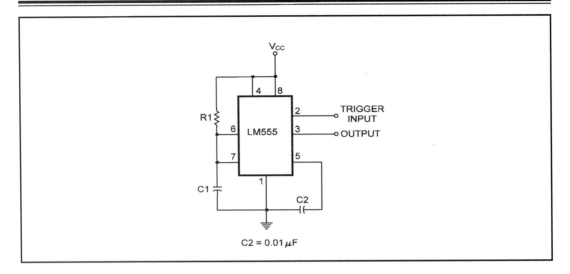

Figure 19-4. LM555 configured as a monostable multivibrator.

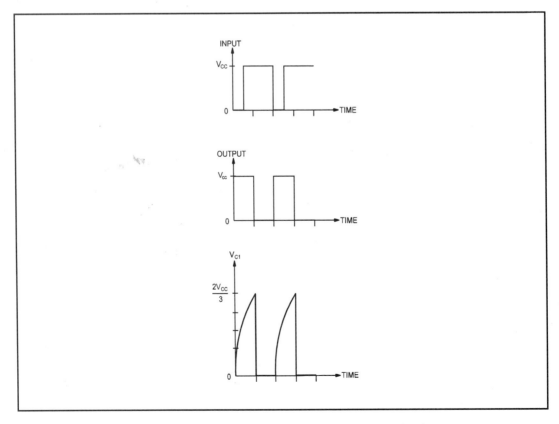

Figure 19-5. Waveforms for the LM555 monostable multivibrator.

LM555 Timer IC

Timing capacitor C1 is initially discharged by the internal discharge transistor. When a negative trigger pulse of less than 1/3 of the supply voltage is applied to the trigger input, the control flip-flop is set. The flip-flop output switches high, and internal transistor Q1 turns off. The timing capacitor starts to charge exponentially toward 2/3 of the supply voltage, for period $T = 1.1 \times R1 \times C1$. The comparator resets the control flip-flop, which in turn discharges the timing capacitor. The output of the LM555 timer IC switches low.

The waveforms for the monostable multivibrator are shown in *Figure 19-5*.

The timing interval is independent of the power supply voltage, because the charge and the threshold levels of the comparator are directly proportional to the power supply voltage. When the output is high during the timing cycle, additional trigger pulses will not affect the circuit. The circuit may be reset by applying a negative pulse to the reset terminal (pin 4) of the LM555 timer. The output will remain in the low state until a trigger pulse is reapplied to the trigger input (pin 2) terminal. To avoid false triggering, the reset terminal (pin 4) of the LM555 timer IC should be connected to the positive rail of the power supply.

A Basic program for designing astable and monostable multivibrators using the LM555 timer is listed in *Table 19-1*. The program asks if either an astable or monostable multivibrator is to be designed. Type "a" or "m," and answer the questions asked by the program. All resistor values are in ohms, all capacitor values are in microfarads, and all times are in seconds. Type "e" to exit the program.

Applications

The monostable multivibrator circuit of *Figure 19-4* can be used as a frequency divider by adjusting the length of the timing cycle.

A pulse-width modulator is a monostable multivibrator that is triggered by a continuous pulse train. The output pulse width can be modulated by applying a signal to pin 5 of the LM555 timer IC. The schematic and waveforms of an LM555 timer pulse-width modulator are shown in *Figure 19-6*.

A pulse-position modulator is an astable multivibrator that is modulated by applying a signal to pin 5 of the LM555 timer IC. The pulse position varies with the modulating

```
100    REM LM555 TIMER CALCULATIONS
110    FOR I = 1 TO 24:PRINT:NEXT I
120    PRINT "LM555 TIMER CALCULATIONS"
130    PRINT:PRINT:PRINT:PRINT
140    INPUT "Astable/Monostable/Exit(a/m/e)?", IN$
150    IF IN$ = "e" THEN END
160    IF IN$ = "m" THEN GOTO 580
170    IF IN$ = "a" THEN GOTO 190
180    GOTO 140
190    REM Do STABLE calculations
200    PRINT
210    INPUT "FREQUENCY?", F
220    PRINT
230    PRINT "0. END"
240    PRINT "1. Find R2, Given R1 = 1000, C1 = USER ENTRY"
250    PRINT "2. Find C1, Given R1 = 1000, R2 = USER ENTRY"
260    PRINT "3. Find C1, Given R1 = USER ENTRY, R2 = USER ENTRY"
270    PRINT "4. Find f, Given C1, R1, R2 USER ENTRIES"
280    PRINT
290    INPUT "Choose:", IN
300    PRINT
310    IF IN = 0 THEN GOTO 140 ELSE ON IN GOSUB 330, 390, 450, 510
330    REM Find R2 given R1 = 1K, C1 ENTERED
340    R1 = 1000
350    INPUT "C1 = ", C1
360    R2 = (1.44/(F * C1/100000) - R1)/2
370    GOSUB 870
380    RETURN
390    REM Find C1 given R1 = 1K, R2 ENTERED
400    R1 = 1000
410    INPUT "R2 = ", R2
420    C1 = 1.44/(F * (R1 + 2 * R2)) * 1000000
430    GOSUB 870
440    RETURN
450    REM Find C1 given R1, R2 ENTERED
460    INPUT "R1 = ", R1
470    INPUT "R2 = ", R2
480    C1 = 1.44/(F * (R1 + 2 * R2)) * 1000000
490    GOSUB 870
500    RETURN
510    REM Find f given C1, R1, R2 ENTERED
520    INPUT "C1 = ", C1
530    INPUT "R1 = ", R1
540    INPUT "R2 = ", R2
550    PRINT
560    PRINT "NEW FREQUENCY WOULD BE: ";1.44/(C1 * 000001 * (R1 + 2 * R2)); "HZ"
570    RETURN
580    REM Do MONOSTABLE calculations
590    PRINT
600    PRINT "0. END"
610    PRINT "1. Find T, given R and C"
```

Table 19-1. Listing for the LM555 timer design program.

```
620     PRINT "2. Find R, given T and C"
630     PRINT "3. Find C, given R and T"
640     PRINT
650     INPUT "Choose: ", IN
660     PRINT
670     IF IN = 0 THEN GOTO 140 ELSE ON IN GOSUB 720, 770, 820
680     PRINT
690     "C = "; C1; "uF"; TAB(15); "R = "; R1; "ohms"; TAB(30); "T = "; T * 1000; "msec"
700     PRINT
710     GOTO 580
720     REM Find T, given R and C
730     INPUT "R = ", R1
740     INPUT "C = ", C1
750     T = 1.1 * R1 * C1 * .000001
760     RETURN
770     REM Find R, given T and C
780     INPUT "T = ", T
790     INPUT "C = ", C1
800     R1 = T/1.1 * C1 * 100000
810     RETURN
820     REM Find C given R and T
830     INPUT "R = ", R1
840     INPUT "T = ", T
850     C = T/1.1 * R1 * .000001
860     RETURN
870     REM Print Astable Variables
880     PRINT
890     PRINT "F = "; F; "Hz"; TAB(15); "C1 = "; C1; "uF"; TAB(30);
900     PRINT "R1 = "; R1; "ohms"; TAB(45); "R2 = "; R2; "ohms"
910     PRINT
920     RETURN
930     END
```

Table 19-1. Listing for the LM555 timer design program (continued).

signal because the threshold voltage at pin 5 is varied. Hence, the time delay is also varied. The schematic and waveforms of an LM555 timer pulse-position modulator are shown in *Figure 19-7*.

A linear ramp generator is shown in *Figure 19-8*. The pull-up resistor, R1, of the monostable multivibrator is replaced by a constant current source. The time interval, T, of the linear ramp is $T = (0.67 V_{cc} R_e (R1 + R2) C1)/(R1 V_{cc} - 0.6(R1 + R2))$

The waveforms of the linear ramp generator are also shown in *Figure 19-8*.

IC Design Projects

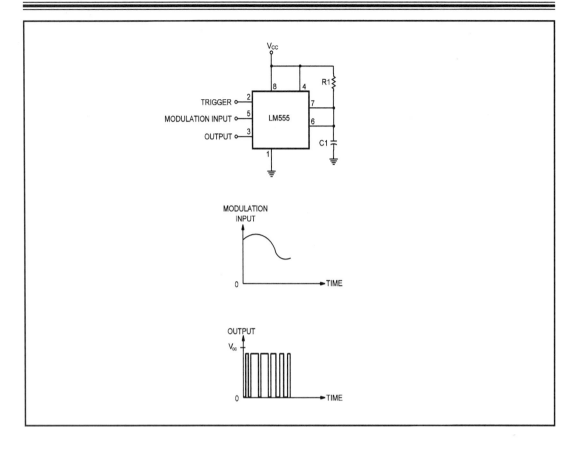

Figure 19-6. Schematic and waveforms of a pulse-width modulator.

The LM555 timer astable multivibrator may be reconfigured into a 50 percent duty cycle oscillator, as shown in *Figure 19-9*. If R2 is less than R1/2, the circuit will not oscillate because the junction of R1 and R2 cannot bring pin 2 down to Vcc/3 and trigger the lower comparator. T1 = 0.693 x R1 x C1 and T2 = [(R1R2)/(R1 + R2)]C1 x Ln[(R2 - 2R1)/2R2 - R1)].

The frequency of oscillation, F, is F = 1/(T1 + T2)

The LM555 timer can be used as a missing pulse detector. The LM555 timer produces an output when an input pulse fails to occur within the delay of the timer. The time delay is set to be slightly longer than the time interval between successive input pulses. The timing cycle is continuously reset by the input pulse train until a change in fre-

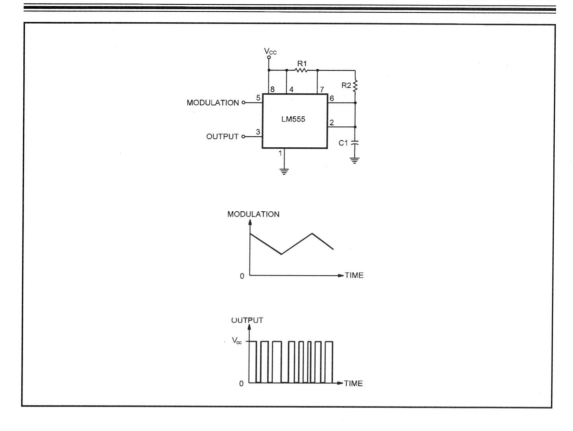

Figure 19-7. Schematic and waveforms of a pulse-position modulator.

quency or a missing pulse allows completion of the timing cycle, which causes a change in the output level.

Problems

Problem 19-1. What are two modes of operation of the LM555 timer IC?
Problem 19-2. In what type of circuit is the LM555 timer used?
Problem 19-3. Why is the LM555 timer classified as a linear IC?
Problem 19-4. Is the output of the LM555 timer a linear output? Why or why not?
Problem 19-5. What is a comparator?
Problem 19-6. What are the threshold and trigger inputs of an LM555 timer?
Problem 19-7. When does a comparator change its output state?
Problem 19-8. What is the function of internal transistor Q1 of an LM555 timer?
Problem 19-9. What drives internal transistor Q1 of an LM555 timer?

IC Design Projects

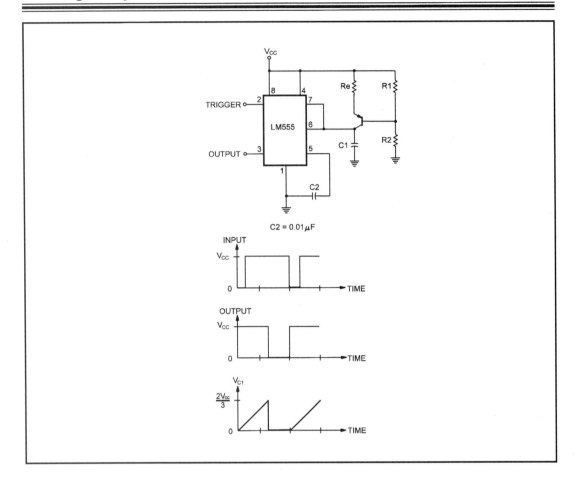

Figure 19-8. Schematic and waveforms of a linear ramp generator.

Problem 19-10. What is the function of internal transistor Q2 of an LM555 timer?

Problem 19-11. What is an astable multivibrator?

Problem 19-12. Calculate the frequency and duty cycle of an LM555 astable multivibrator if R1 = 1k, R2 = 100k and C1 = 0.01uF.

Problem 19-13. Calculate the minimum R1 that can be used in an LM555 timer astable multivibrator if Vcc = 15 volts.

Problem 19-14. What is a monostable multivibrator?

Problem 19-15. Calculate the pulse width of an LM555 monostable multivibrator if R1 = 47k and C1 = 0.01 uF.

Problem 19-16. How is false triggering avoided in LM555 timer circuits?

Problem 19-17. How can an LM555 monostable be changed into a pulse-width modulator?

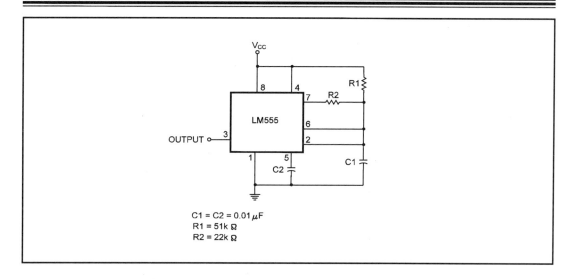

Figure 19-9. Schematic of a 50 percent duty cycle oscillator.

Problem 19-18. How can an LM555 astable be changed into a pulse-position modulator?

Problem 19-19. How can an LM555 monostable be changed into a linear ramp generator?

Problem 19-20. Calculate the time interval of an LM555 linear ramp generator if R1 = 1k, R2 = 4.7k, Re = 100 ohms, C1 = 0.1 uF and Vcc = 10 volts.

Problem 19-21. What condition must be met for an LM555 timer 50-percent duty cycle oscillator to oscillate?

Chapter 20
◆ DC Motor ◆
◆ Speed Controller ◆

The speed of a DC motor can be varied by changing either the field current or the armature voltage. Normally, the field current is controlled by a series-connected rheostat.

A rheostat is a variable resistor. This method of speed control is inefficient because there are high power losses in the copper wire windings of the rheostat. The DC motor characteristics are adversely affected when the speed is controlled by varying a rheostat connected in series with the DC motor field winding.

The preferred method of changing the DC motor speed is by varying the armature voltage. The load characteristics are similar to those at base speed. The base speed is the speed of the DC motor with the maximum field current and the rated armature voltage applied.

The DC motor drive consists of a chopper that uses an electronic switch interposed between the DC source and the DC motor. By varying the duty cycle of the electronic switch, the average armature voltage (and therefore the speed) of the DC motor is efficiently controlled.

At low power (less than 50 kilowatts), the transistor chopper is the most economical means of speed control. At higher power (greater than 50 kilowatts), thyristors must be used. The chopper speed control circuit is used mainly in the traction control of subways, locomotives and electric vehicles.

Circuit Operation

The DC motor speed controller is an LM555 timer astable multivibrator with a variable duty cycle. The schematic of the DC motor speed control is shown in *Figure 20-1*. Timing capacitor C1 charges through resistor R2, diode D2 and the right-hand

IC Design Projects

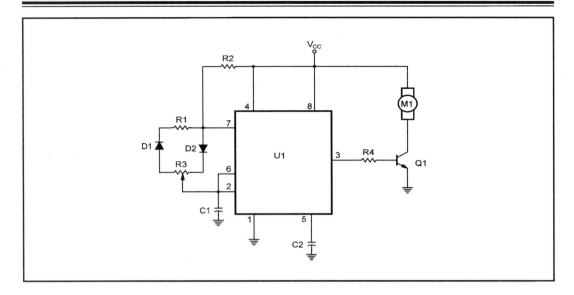

Figure 20-1. Schematic of the DC motor speed controller.

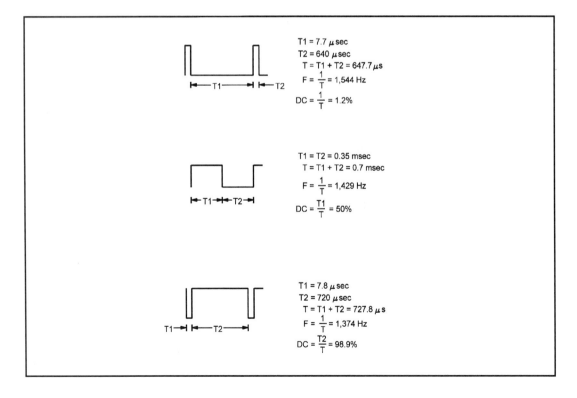

Figure 20-2. Waveforms for the DC motor speed controller.

DC Motor Speed Controller

portion of potentiometer R3. Timing capacitor C1 discharges through the left-hand portion of potentiometer R3, diode D1 and resistor R1.

The LM555 timer astable multivibrator, U1, oscillates at about 1400 Hertz. The waveforms at U1 pin 3 for both the middle and extreme positions of potentiometer R3 are shown in *Figure 20-2*. The duty cycle is adjustable from about one percent to 99 percent.

The output of the LM555 timer astable multivibrator is fed to an inverter switch consisting of resistor R4 and transistor Q1. Resistor R4 limits the base current flow into transistor Q1. Transistor Q1 must be able to handle the current requirements of the DC motor. The 2N3055 transistor can handle up to 15 amperes of current if the transistor is adequately heat sunk. Transistor Q1 is a power transistor. It enables the speed control to drive relatively heavy loads.

The power supply voltage for the DC motor controller must be in the range of 5 volts to 15 volts. If the DC motor requires a higher DC voltage, a voltage regulator can be used to provide the required 5 to 15 volts to the LM555 timer IC.

Construction

The DC motor speed controller may be built on a piece of perfboard. *Table 20-1* is the parts list for the DC motor speed controller. The power transistor, IC and diodes must be properly oriented or damage will result. The power transistor should be mounted on an adequately-sized heat sink. The specified power transistor can dissipate over 100 watts when mounted on a proper heat sink.

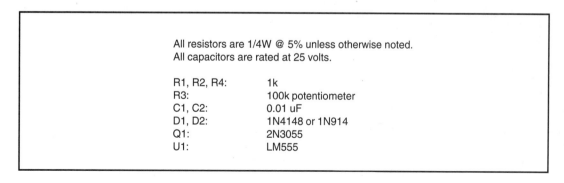

All resistors are 1/4W @ 5% unless otherwise noted.
All capacitors are rated at 25 volts.

R1, R2, R4:	1k
R3:	100k potentiometer
C1, C2:	0.01 uF
D1, D2:	1N4148 or 1N914
Q1:	2N3055
U1:	LM555

Table 20-1. Parts list for the DC motor speed controller.

If a metal enclosure is used, it can serve as the heat sink for the power transistor. The power transistor must be electrically (not thermally) isolated from the heat sink by using mica washers and a suitable transistor socket.

Most problems are caused by cold solder joints and short circuits created by solder bridges. If you are inexperienced with soldering, you should practice soldering on a scrap piece of perfboard.

Other Uses

The DC motor speed controller can be used in other applications. If the load is a low-voltage incandescent bulb, then the circuit becomes a lamp dimmer. If the load is an infrared-emitting diode, then the circuit can be used as a remote control transmitter. In this application, potentiometer R3 should be replaced by two equal resistors whose total resistance is equal to that of the potentiometer.

Chapter 21
♦ Electronic Organs ♦

A monophonic organ is an organ in which only one note at a time may be played. The first electronic organ project listed here is a two-octave monophonic organ.

A polyphonic organ is an organ in which several notes may be played simultaneously. Chords can be played on a polyphonic organ. The second electronic organ project is a two-octave polyphonic organ.

Both organs cover the musical range from C (262 Hz) to C (1047 Hz), with twelve notes per octave. This includes sharps and flats. *Table 21-1* lists the notes, frequencies and time periods for each of the twenty-five musical notes that are available on both electronic organs. Both electronic organs feature tremolo and pitch control circuits.

NOTE	FREQUENCY (HZ)	TIME PERIOD (MILLISEC.)
C	262	3.82
C#	277	3.61
D	294	3.40
D#	311	3.22
E	330	3.03
F	349	2.87
F#	370	2.70
G	392	2.55
G#	415	2.41
A	440	2.27
A#	466	2.15
B	494	2.02
C	523	1.91
C#	554	1.81
D	587	1.70
D#	622	1.61
E	659	1.52
F	698	1.43
F#	740	1.35
G	784	1.28
G#	831	1.20
A	880	1.14
A#	932	1.07
B	988	1.01
C	1047	0.96

Table 21-1. Notes available on both electronic organs.

IC Design Projects

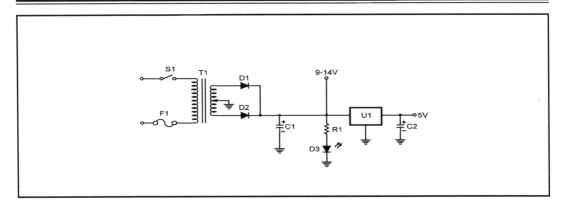

Figure 21-1. Power supply for both organs.

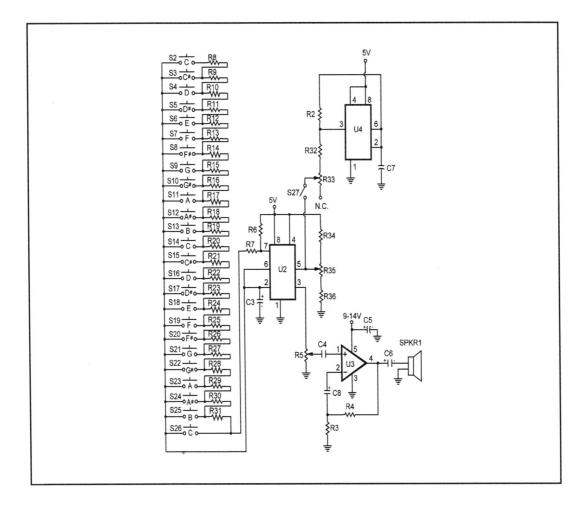

Figure 21-2. Electronic organ for playing one note at a time.

Electronic Organs

Circuit Description

The power supply used for both electronic organs is shown in *Figure 21-1*. Transformer T1 is a center-tapped step-down transformer. Diodes D1 and D2 rectify the low AC voltage.

Capacitor C1 smooths the pulsating DC voltage which is used to power the audio amplifier. Resistor R1 limits the current flow through light-emitting diode D3 which functions as a "power-on" indicator. The output of the three-terminal voltage regulator U1 is 5 volts. The 5 volts powers the oscillator and tremolo circuits. Capacitor C2 improves the transient response of the three-terminal voltage regulator, U1.

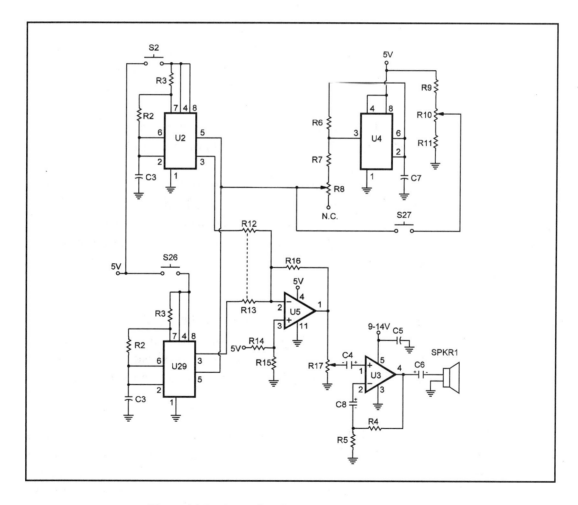

Figure 21-3. Organ for playing several notes at a time.

IC Design Projects

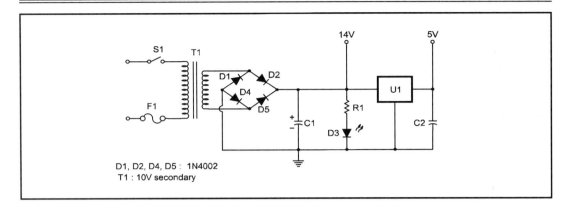

Figure 21-4. Alternate power supply for both electronic organs.

The monophonic electronic organ is shown in *Figure 21-2*. The frequency of oscillation of the astable multivibrator U2 is determined by the timing capacitor, C3, and the timing resistor(s) selected by switches S2 - S26. These normally-open push-button switches can select any resistor(s) in the resistor chain, R7 - R31. Resistors R7 - R31. have been selected for the closest possible approximation of the musical notes, within one or two Hertz. You may choose to replace each resistor in the resistor chain, of the next lowest standard value resistor in series, with a 1000-ohm trimmer potentiometer.

The output of U2 is fed via volume control R5 to the input of U3, which is a TDA-2002 audio amplifier. The TDA-2002 has internal thermal overload and short-circuit protection circuits. The frequency response should exceed 40 Hz to 15 kHz within three decibels. The maximum power supply voltage is 14 volts. This power amplifier can deliver at least 5 watts into a load impedance of 2 to 16 ohms, with less than 5 percent harmonic distortion.

The astable multivibrators, U2 and U4, are powered by the 5-volt supply, while the audio amplifier, U3, is powered by the 9-volt to 14-volt power supply. If the astable multivibrators are powered by the higher voltage, the output of astable multivibrator U2 would saturate audio amplifier U3, which is configured as a noninverting amplifier with a voltage gain of 2.

The tremolo circuit consists of U4 and its associated components, and is configured as an astable multivibrator. The tremolo circuit oscillates at about 5 Hertz. Pin 6 is connected to pin 3 (output) via resistor R2, instead of pin 7 (discharge), allowing the trigger and threshold inputs of U4 to float rather than being tied to the power supply via a pull-up resistor. The discharge and output of the LM555 timer serve similar functions. The tremolo can be varied with tremolo control R33 and switch S27. If S27 is open, there is no tremolo because the tremolo circuit no longer modulates U2. When the tremolo is active, the pitch of the organ can be adjusted by pitch control R35.

The polyphonic electronic organ is shown in *Figure 21-3*. Its operation is similar to the monophonic organ. The main difference is that each musical note is generated by its own LM555 astable multivibrator (U2, U6 - U29) and their associated components. Timing capacitor C3 is the same for all of the musical notes. However, timing resistor R2 is different for each of the musical notes. For each R2, you may want to substitute the next-lowest standard value resistor in series with a trimmer potentiometer. The outputs of U2 and U6 - U29 are fed to U5, which is configured as a unity gain summing amplifier. The tremolo circuit consists of U4 and its associated components. The audio amplifier consists of U3 and its associated components.

You can power the musical note generator circuits, tremolo and mixer circuits with the power supply used to power the audio amplifier if (and only if) R16 is changed from a 10,000-ohm resistor to a 1000-ohm resistor. The audio amplifier will not be saturated by the output of the summing amplifier because it now operates as a summing attenuator with an attenuation factor of 10.

If you have a doorbell transformer with a secondary rating of 10 volts, the power supply shown in *Figure 21-4* may be used to power either electronic organ. The rectifier bridge consists of diodes D1, D2, D4 and D5.

Construction

The electronic organs may be built on one or more pieces of perfboard. The author used one piece of perfboard for the timing resistor(s) chain and a second piece of perfboard for the rest of the monophonic organ. The power supply was built on a third piece of perfboard.

The timing resistors, R7 - R31, may each be substituted with the next-lowest standard value resistor in series with a 1000-ohm trimmer potentiometer. The frequency of each musical note of the monophonic organ may therefore be accurately tuned by adjusting the appropriate trimmer potentiometer.

IC Design Projects

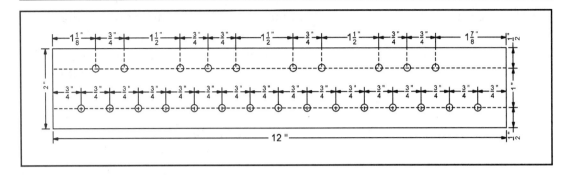

Figure 21-5. Template for keyboard, drawn one-half scale.

An adequate heat sink should be fabricated for the audio amplifier. The heat sink can be made of steel or aluminum, and it should be electrically (not thermally) isolated from the audio amplifier IC, U3.

Care should be taken to avoid cold solder joints and solder bridges. All diodes, capacitors and ICs must be properly oriented.

The cabinet design is limited only by your imagination. The author used 25 normally-open SPST push-button switches for the keyboard. A drilling template for the keyboard is shown in *Figure 21-5*. The holes drilled should be approximately 1/4" in diameter, to suit the push-buttons used.

The parts list for the monophonic electronic organ is given in *Table 21-2*. The parts lists for the polyphonic organ are given in *Table 21-3* and *Table 21-4*.

If you build the polyphonic electronic organ, you need several pieces of perfboard to build the 25 musical-note-generating astable multivibrators. The values of R2 for the polyphonic organ are given in *Table 21-4*. The next-lowest standard value resistor in series with a trimmer potentiometer may be substituted for R2. The trimmer potentiometer is used to tune each musical note. From C (1047 Hz) to G# (831), a 1000-ohm trimmer potentiometer may be used. From G (784) to B (494), a 5000-ohm trimmer potentiometer should be used. From A# (466) to C (262), a 10,000-ohm trimmer potentiometer may be used.

The monophonic organ may be expanded to three or more octaves in two ways. The timing resistor chain may be extended to include several additional musical notes.

Alternately, a second monophonic organ may be built with the resistors of the timing resistor chain selected to the next two lower or upper octaves.

The polyphonic organ may also be expanded to three or more octaves, when an astable multivibrator is built and tuned to each additional musical note. The output of each additional astable multivibrator is fed via a 10,000-ohm resistor to the summing amplifier, U5, as shown in *Figure 21-3*.

All resistors are 1/4W @ 5% unless otherwise noted.
All capacitors are rated at 25 volts or greater.

R1, R22, R25-R27:	1k
R2:	68k
R3, R4:	220
R5, R33:	10k potentiometer
R6:	10k
R7:	7.5k
R8, R9:	2.7k
R10, R11:	2.4k
R12, R14:	2k
R13, R34:	2.2k
R15-R17:	1.8k
R18-R21, R23:	1.5k
R24:	1.2k
R28, R29:	820
R30, R31:	680
R32:	3.3k
R35:	5k potentiometer
R36:	6.8k
C1:	4700 uF
C2:	0.1 uF
C3:	56 nF
C4:	10 uF
C5:	68 nF
C6:	1000 uF
C7:	1 uF
C8:	100 uF
U1:	LM7805
U2, U4:	LM555
U3:	TDA-2002 or LM383
D1, D2:	1N4001
D3:	LED
S1, S27:	SPST
S2-S26:	N.O. SPST
T1:	12.6 VCT secondary
F1:	1/2A slow blow
SPKR1:	4-8 ohm speaker

Table 21-2. Parts list for the monophonic electronic organ.

All resistors are 1/4W @ 5% unless otherwise noted.
All capacitors are rated at 25 volts or greater.

R1:	1k
R2:	See Table 16-4
R3:	10k
R4, R5:	220
R6:	68k
R7:	3.3k
R8, R17:	10k potentiometer
R9:	2.2k
R10:	5k trimmer potentiometer
R11:	6.8k
R12-R16:	10k
U2, U4, U6-U29:	LM555
U5:	LM324
S1, S27:	SPST
F1:	1/2A slow blow
C1:	4700 uF
C2:	0.1uF
C3:	56 nF
C4:	10 uF
C5:	68 nF
C6:	1000 uF
C7:	1 uF
C8:	100 uF
D1, D2:	1N4001
D3:	LED
U1:	LM7805
U3:	TDA-2002 or LM383
T1:	12.6 V.C.T. secondary
S2-S26:	N.O. SPST
SPKR1:	4-8 ohm speaker

Table 21-3. Parts list for the polyphonic electronic organ.

Tuning and Use

If the electronic organ has been built using only fixed-value resistors, no tuning is required. Simply press the switches of the keyboard and play music.

If either electronic organ has been built using trimmer potentiometers, the organ must be tuned. Switch S27 must be in the open position. Each trimmer potentiometer is adjusted until the output of the appropriate astable multivibrator is correct, according

Electronic Organs

to *Table 21-1*. If a frequency counter is used to tune the organ, use the frequency data of *Table 21-1*. If an oscilloscope is used to tune the organ, use the time period data of *Table 21-1*. The organ is now ready to be played.

For tremolo, set S27 in the closed position. The tremolo and pitch controls can be adjusted for the desired sound effects.

NOTE	FREQUENCY (Hz)	R2 (OHM) OF FIGURE 21-3
C	1047	7.5K
B	988	8.2K
A#	932	8.9K
A	880	9.7K
G#	831	10.5K
G	784	11.5K
F#	740	12.5K
F	698	13.5K
E	659	14.7K
D#	622	16.2K
D	587	17.2K
C#	554	18.7K
C	523	20.2K
B	494	21.7K
A#	466	23.2K
A	440	25.0K
G#	415	26.8K
G	392	28.6K
F#	370	30.6K
F	349	30.8K
E	330	32.8K
D#	311	35.2K
D	294	37.6K
C#	277	40.3K
C	262	43.0K

Table 21-4. Resistor R2 values for the polyphonic organ.

Chapter 22
◆ Automatic Light Timer ◆

When your house is dark every night for a few nights, thieves suspect that you are away on holiday. This is an invitation for them to break in and help themselves to your possessions.

The automatic light timer turns up to 10 lights on and off sequentially, night after night, to give the impression that you are home. The sequential operation gives the impression that you are leaving one room and entering another room.

In the event of a power failure, the automatic light timer features a backup battery to ensure that the automatic light timer never loses time. When the AC power is restored, the lights continue to turn on and off at the proper times. There is no programming required for the automatic light timer. Simply plug it in.

Circuit Description

The automatic light timer circuit is shown in *Figure 22-1*. IC U1 is an LM555 timer configured as an astable multivibrator. Resistors R1 and R3, potentiometer R2 and capacitor C1 are the external timing components for the LM555 timer IC. Remember that each day contains 86,400 seconds.

ICs U2 - U4 are CD4017 CMOS decade counters. The pin assignment for the CD4017 IC is shown in *Figure 22-2*. ICs U2 - U4, in combination, are configured as a divide-by-1000 counter.

The CD4017 consists of a five-stage Johnson decade counter (decade shift counter) and an output decoder. The output decoder converts the Johnson binary code to a decimal number.

The decade counter advances one count at the leading edge of the clock pulse when the clock enable and reset signals are low. The decoded outputs are spike-free, and the proper count sequence is assured with antilock gating. The ten decoded outputs are normally low; they are high only at their respective decimal time slot for one full clock

IC Design Projects

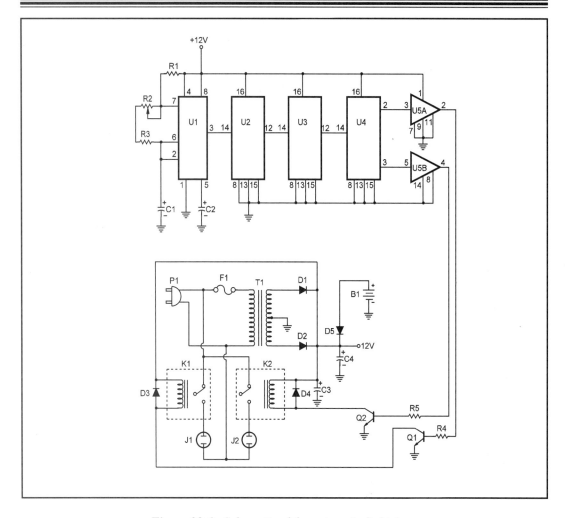

Figure 22-1. Schematic of the automatic light timer.

pulse cycle. The carry-out signal completes one cycle for every ten clock pulses, as shown in *Figure 22-3*. The carry-out signal is used as the clock pulse for the succeeding decade counter in a multi-decade counter chain. The timing diagram of the CD4017 is shown in *Figure 22-3*. The "0" to "9" outputs of IC U4 are high for 2.4 hours, once every 24 hours.

IC U5 is a CD4050 CMOS hex buffer. Its pin assignment is also shown in *Figure 22-2*. IC U5 is a buffer/driver, used as an interface between the CMOS decade counter, U4, and the transistor inverter switches.

Automatic Light Timer

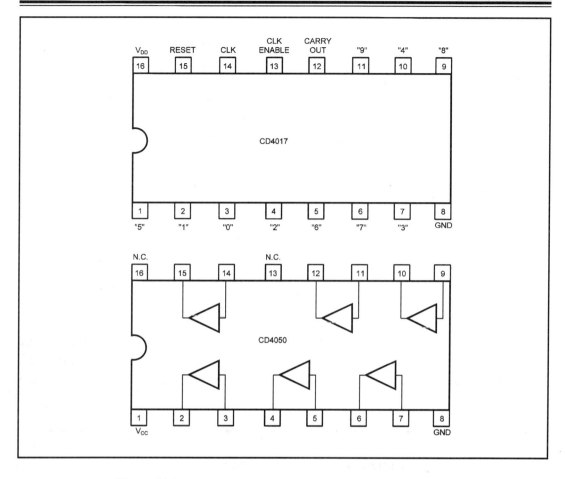

Figure 22-2. Pin assignments for the CD4017 and CD4050 ICs.

Resistors R4 and R5 limit the base current drive to transistors Q1 and Q2, respectively. The transistor switches control relays K1 and K2. Diodes D3 and D4 protect the transistor switches from inductive spike voltages generated by the relays when they are energized. When an output pulse of U4 is positive, there is sufficient base drive to saturate the appropriate transistor switch. When the transistor switch is saturated, its collector goes to about 0.1 volts and the relay is energized, closing the relay contacts. Household AC line voltage is connected to the appropriate AC socket. The relay coil impedance must be at least 100 ohms.

Transformer T1 steps down the household AC line voltage to a lower AC voltage. Diodes D1 and D2 rectify the low AC voltage into a pulsating DC voltage. Capacitors C3 and C4 smooth the pulsating DC voltage. The lamps to be controlled are plugged into AC sockets J1 and J2.

IC Design Projects

In the event of a power failure, backup battery B1 supplies power to the ICs to maintain correct timing. The relays will not be energized during a power failure. Diode D5 blocks current flow to the battery when there is AC power.

Although only two lamp control-switch circuits are shown in *Figure 22-1*, up to 10 lamp control-switch circuits may be controlled by IC U4. Each output, "2" to "9," can be used to control a lamp by using an additional control-switch circuit consisting of a buffer, a transistor switch, a relay and associated components. When one lamp is turned off, another lamp is turned on.

Construction

The automatic light timer can be built on a piece of perfboard. The parts list is given in *Table 22-1*. IC sockets should be used. All ICs, diodes, capacitors and transistors must be properly oriented. You should also use relay sockets.

Take care to avoid cold solder joints and short circuits created by solder bridges. If you are inexperienced with soldering, you should practice on a scrap piece of perfboard. Most problems are caused by cold solder joints and by short circuits created by solder bridges.

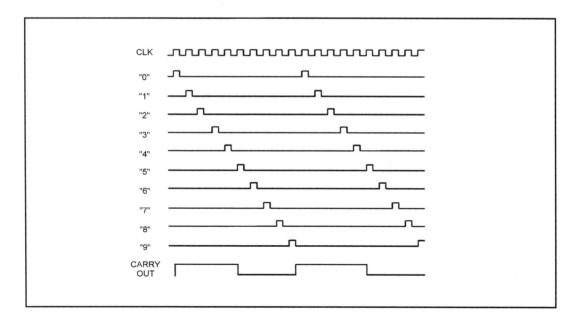

Figure 22-3. Timing diagram for the CD4017 IC.

Automatic Light Timer

Any plastic or metal enclosure can be used. Machine holes for the AC sockets and the AC cord grommet. A hole is also required for the fuse holder if a chassis-mount fuse holder is to be used. An in-line fuse may also be used; however, if it blows, it must be desoldered from the circuit, and a replacement fuse must be resoldered into the circuit.

Calibration and Use

Calibration of the automatic light timer is easy. A wristwatch with a second hand is required for calibrating. Connect the positive test lead of a DC voltmeter to U1 pin 3. Connect the negative test lead of the DC voltmeter to ground. Adjust trimmer potentiometer R2 until the output of U1 has a clock pulse of about 86 seconds (86,400/1000). That is, U1 pin 3 should be high for about 43 seconds, and U1 pin 3 should be low for about 43 seconds. It may be difficult to adjust trimmer potentiometer R2 for an 86.4 second clock pulse at U1 pin 3.

A clock pulse in the range of 85 seconds to 87 seconds is sufficient for the lamps to turn on and off within thirty minutes (earlier or later) of the previous day's on and off times. This provides a "randomness" to the on and off times, so it looks as if you are home.

All resistors are 1/4W @ 5% unless otherwise noted.
All capacitors are rated at 25 volts.

R1:	1k
R2:	500k potentiometer
R3:	750k
R4, R5:	10k
C1:	68 uF tantalum
C2:	0.01 uF
C3, C4:	1000 uF
U1:	LM555
U2-U4:	CD4017
U5:	CD4050
Q1, Q2:	2N2222A
D1, D2:	1N4002
D3, D4:	1N4148 or 1N914
K1, K2:	SPDT 12-volt DC relay
J1, J2:	AC sockets
T1:	18 V.C.T. secondary at ½ ampere
F1:	½ ampere slow blow
P1:	AC plug with line cord

Table 22-1. Parts list for the automatic light timer.

The automatic light timer is easy to use. Simply plug it in about 24 hours before the time that the first lamp is to turn off. You can experiment with the plug-in times until you are familiar with the timing of the automatic light timer. The automatic light timer should provide years of reliable service.

Chapter 23
◆ Warble Alarm ◆

The warble alarm produces a two-tone sound similar to that of European emergency vehicles. The alarm can be used in many applications in which a warning sound is required. The warble alarm can be powered by a 9-volt battery, a car battery, or by a 9-volt to 15-volt AC operated power supply.

The alarm is easy to build and use. It can be housed in any suitable enclosure. The small size of the warble alarm allows it to be hidden from view, a useful feature for security systems. The two-tone sound of the warble alarm should grab the attention of any would-be thief.

Circuit Description

Two LM555 timer ICs are configured as astable multivibrators, as shown in *Figure 23-1*. IC U1 is configured as a free-running oscillator operating at 1 Hertz. IC U2 is also

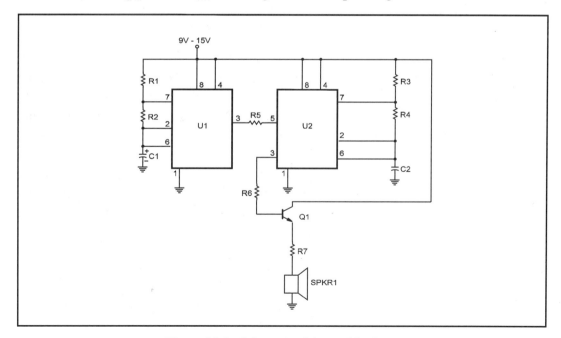

Figure 23-1. Schematic of the warble alarm.

IC Design Projects

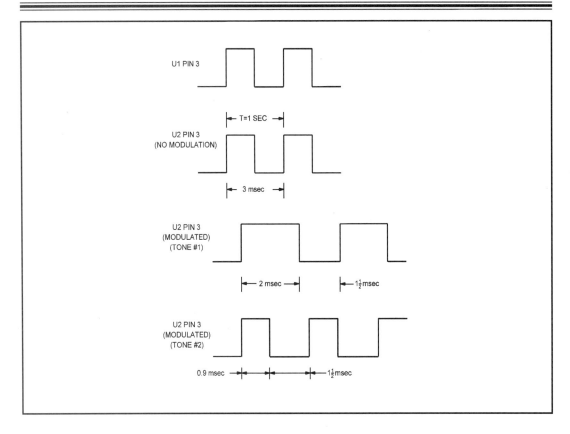

Figure 23-2. Waveforms of the warble alarm.

configured as a free-running oscillator. It oscillates at 330 Hertz when it is not being modulated by U1. The output of the U1 frequency modulates U2 through the current-limiting resistor, R5. The output frequency of IC U2 alternates between 285 Hertz and 420 Hertz in one-second intervals.

The output waveforms are shown in *Figure 23-2*. The top two waveforms are the output signals measured at U1 pin 3 and U2 pin 3, when U2 is not being modulated by U1. The bottom two waveforms are the two alternating output signals measured at U2 pin 3 when U2 is being modulated by U1.

Transistor Q1 drives the 8-ohm speaker. Resistor R6 limits the base current flowing into transistor Q1. Resistor R7 limits the current flow through the speaker coil; therefore, R7 can be used to adjust the acoustical output of the speaker. Resistor R7's values and power ratings are listed in *Table 23-1*.

Warble Alarm

R7 (ohms)	Power Rating (Watts)
47	2
82	1
100	1

Table 23-1. Resistor R7 values and power ratings.

The power supply voltage should be in the range of 9 volts to 15 volts. The warble alarm can be battery powered, or it can be operated from an AC power supply. If it is to be battery powered, resistor R7 should be at least 120 ohms to minimize the battery current drain.

Construction

The warble alarm is an easy project to build, as can be seen from the parts list given in *Table 23-2*. It can be built on a piece of perfboard. The ICs, transistor and tantalum capacitor must be properly oriented. IC sockets should be used to prevent heat damage to the ICs caused by soldering.

Care should be taken when soldering the circuit. Most problems are caused by cold solder joints or by short circuits created by solder bridges. If you have little experience with soldering, you should practice soldering on a scrap piece of perfboard.

All resistors are 1/4W @ 5% unless otherwise indicated.
All capacitors are rated at 25 volts.

R1, R3, R5:	10k
R2:	75k
R4:	220k
R6:	2.7k
R7:	100 ohms at 1W; see text
SPKR1:	8 ohms
U1, U2:	LM555
Q1:	2N3055
C1:	10 uF tantalum
C2:	0.01 uF

Table 23-2. Parts list for the warble alarm.

Resistor R5 should not be installed until you are instructed to do so in the test procedure.

Test Procedure

Connect the positive test lead of a DC voltmeter to IC U1 pin 3. Connect the negative test lead of the DC voltmeter to ground. A 1-Hertz square wave should be present, and the needle of the DC voltmeter should move back and forth once each second.

A steady tone of 330 Hertz should be heard from the loudspeaker. Turn off the power to the warble oscillator. Connect resistor R5 into the circuit. Reapply power to the warble alarm. A two-tone sound should be heard from the speaker.

Chapter 24
◆ Darkroom Timer ◆

The darkroom timer has a normally-off AC socket for the enlarger and a normally-on AC socket for the safe light. The darkroom timer features a timer/off/focus switch.

In the focus position, power is only applied to the relay coil. The enlarger can be focused in this mode. In the timer position, power is applied to the LM555 timer IC and associated components. The circuit operates as a timer in this mode.

The darkroom timer also features a short/long switch. In the short position, the potentiometer can be used to adjust the timing interval from 1 second to 12 seconds. In the long position, the potentiometer can be used to adjust the timing interval from 11 seconds to 120 seconds.

Circuit Operation

The darkroom timer incorporates relay load switching as well as a switchable and adjustable timing delay. The schematic of the darkroom timer is shown in *Figure 24-1*. The timer can be adapted to other applications.

Transformer T1 steps down the household AC line voltage to a lower AC voltage. Diodes D1 and D2 rectify the low AC voltage. Capacitor C1 smooths the pulsating DC voltage into a fluctuating DC voltage. Voltage regulator U1 maintains the darkroom timer supply voltage at a constant 12 volts. Capacitor C3 improves the transient response of U1. Switch S1 is the timer/off/focus switch. In the focus position, power is applied only to the relay coil. The enlarger can be focused, and light-emitting diode (LED) D6 illuminates. In the timer position, power is applied to the LM555 and associated components. The circuit operates as a timer, and LED D5 illuminates. Resistors R4 and R5 limit the current flow through LEDs D5 and D6, respectively.

IC U2 is an LM555 timer configured as a monostable multivibrator. Each time that power is applied to the circuit (capacitor C5 is initially discharged), capacitor C5 charges up from 0 volts to 12 volts through resistor R1. This sequence generates a trigger pulse at U2 pin 2. A single pulse is generated at U2 pin 3. The pulse width is determined by

IC Design Projects

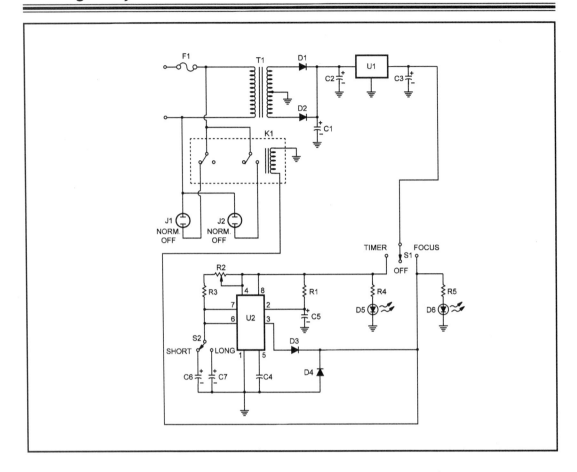

Figure 24-1. Schematic of the darkroom timer.

timing components R2, R3 and C6 or C7. Switch S2 is used to select short or long pulse width. Potentiometer R2 is used to adjust the pulse width. In the short mode, the pulse width can be varied from one second to twelve seconds. In the long mode, the pulse width can be varied from 11 seconds to 120 seconds.

The LM555 timer IC powers the relay coil. The impedance of the relay coil must be at least 100 ohms. Diodes D3 and D4 protect U2 from inductive-switching damage caused by inductive spike voltages generated by the relay when it is switched on.

AC socket J1 is normally on, and AC socket J2 is normally off. The safe light is plugged into the normally-on AC socket, and the enlarger is plugged into the normally-off AC socket. Fuse F1 protects the darkroom timer, safe light and enlarger.

Darkroom Timer

All resistors are 1/4W @ 5% unless otherwise noted.
All capacitors are rated at 25 volts unless otherwise noted.

R1:	10k
R2:	1M linear potentiometer
R3:	100k
R4, R5:	2.2k
C1:	1000 uF
C2:	0.33 uF tantalum
C3:	1.0 uF tantalum
C4:	0.01 uF
C5:	47 uF
C6:	10 uF tantalum
C7:	100 uF 16-volt tantalum
D1, D2:	1N4001
D3, D4:	1N914 or 1N4148
D5, D6:	Light-emitting diode
U1:	LM7812
U2:	LM555
S1:	SP3T
S2:	SP2T
J1, J2:	AC sockets
K1:	DPDT 12-volt DC relay
T1:	25.2 V.C.T. secondary at ½ ampere
F1:	2A slow blow

Table 24-1. Parts list for the darkroom timer.

Construction

The darkroom timer can be built on a piece of perfboard. The ICs, diodes and capacitors must be properly oriented. It is recommended that IC sockets are used. The parts list for the timer is given in *Table 24-1*.

Most problems are caused by cold solder joints and by short circuits created by solder bridges. If you are inexperienced with soldering, you should practice on a scrap piece of perfboard.

The darkroom timer can be housed in a plastic or metal enclosure. Machine holes for the AC sockets, LEDs, potentiometer, fuse holder, switches and AC line cord grommet, as well as the mounting holes for the transformer and the piece of perfboard on which the darkroom timer is built.

Testing and Use

Plug the darkroom timer into an AC socket. Place switch S2 into its short position. With switch S1 in the focus position, there should be 12 volts across the relay coil and LED D6 should illuminate. If not, verify the wiring of the power supply and of switch S1. Verify the orientation of LED D6.

With switch S1 in the timer position, LED D5 should illuminate and the relay should "click" on and off once for 1 to 11 seconds. If not, verify the wiring of IC U2, its associated components, and switch S2. Verify the orientation of LED D5.

Place switch S2 in its long position, LED D5 should illuminate and the relay should "click" on and off once for 11 to 120 seconds. If not, verify the wiring of IC U2, its associated components, and switch S2.

The safe light is plugged into the normally-on AC socket, and the enlarger is plugged into the normally-off AC socket. The desired exposure time is obtained by placing switch S2 into the appropriate position and by adjusting potentiometer R2.

To focus the enlarger, place switch S1 in the focus position. To expose the photographic paper, place switch S1 in the timer position and adjust the darkroom timer for the correct exposure time. The darkroom timer should provide years of reliable service.

Part Six
♦ LM567 Tone ♦
♦ Decoder IC ♦

Chapter 25
◆ LM567 Tone Decoder IC ◆

The LM567 tone decoder provides a low output when an input signal within the passband is present. The schematic and pin assignment for the dual-in-line package are shown in *Figure 25-1*.

The LM567 tone decoder contains an I-phase detector, a VCO, a Q-phase detector, and a transistor switch. The I-phase detector and the Q-phase detector are driven by the VCO. The VCO determines the center frequency of the LM567 tone decoder. External components are used independently to set the center frequency, bandwidth and output delay of the LM567 circuit.

The LM567 tone decoder can be set to detect any input frequency between 0.01 Hertz and 500 kilohertz. When the input frequency is very low, it may take the LM567 tone decoder one second or more to lock-on to the input signal.

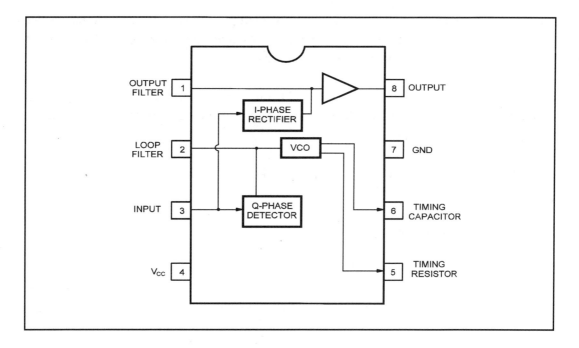

Figure 25-1. Schematic and pin assignment of the LM567 IC.

IC Design Projects

The LM567 tone decoder has several features. It has a 20-to-1 frequency range with an external resistor. The LM567 has a logic compatible output which can sink 100 milliamperes of current. Its bandwidth is adjustable from 0 percent to 14 percent. The LM567 rejects out of band signals and noise. It is immune to false signals. The LM567 has a highly stable center frequency which is adjustable from 0.01 Hertz to 500 kilohertz.

Pin 1 is connected to the output of the I-phase detector. It can be connected to an external capacitor if a low-pass output filter is required.

Pin 2 is connected to the output of the Q-phase detector and the output of the VCO. It can be connected to an external capacitor if a low-pass loop filter is required.

The capacitor connected to pin 1 of the LM567 tone decoder should have a greater capacitance than the capacitor connected to pin 2 of the LM567 tone decoder IC.

Pin 3 is connected to the input of the I-phase detector and the input of the Q-phase detector. The input signal is connected to pin 3 of the LM567 tone decoder IC. A coupling capacitor should be used to remove the DC component of the input signal.

Pin 4 is connected to the positive rail of a power supply whose voltage must be greater than 4.75 volts but less than 9 volts. A decoupling capacitor can be connected between the Vcc (pin 4) and the ground (pin 7) terminals.

Pin 5 is an output of the VCO. The external timing resistor is connected between pin 5 and pin 6 of the LM567 tone decoder. The external timing resistor should be between 2 kilohms and 20 kilohms.

Pin 6 is also an output of the VCO. The external timing capacitor is connected between pin 6 and ground (pin 7) of the LM567 tone decoder IC.

Pin 7 is connected to ground.

Pin 8 is the output of the LM567 tone decoder. Pin 8 goes low when the input signal frequency (signal at pin 3) matches the center frequency of the LM567 tone decoder. A pull-up resistor is connected between pin 8 and Vcc (pin 4) of the LM567 tone decoder IC.

LM567 Tone Decoder IC

Applications

The LM567 tone decoder can be used in many applications. They include touch-tone decoding, oscillators, frequency monitoring and control, wide-band FSK demodulation, ultrasonic controls, carrier-current remote control and communications paging decoders.

The schematic of a quadrature (or two-phase) oscillator is shown in *Figure 25-2*. The input terminal (pin 3) is connected to ground. The output terminal (pin 8) and the timing resistor terminal (pin 5) are the two output terminals of the quadrature oscillator. The two outputs of a quadrature oscillator are 90 degrees out of phase with each other. The frequency of oscillation, F, is $F = 1/R1C1$.

The schematic of a two-frequency oscillator is shown in *Figure 25-3*. The input terminal (pin 3) is connected to the timing resistor terminal (pin 5). The output terminal (pin 8) and the timing resistor terminal (pin 5) are the two output terminals of the two-frequency oscillator.

The schematic of a tone decoder is shown in *Figure 25-4*. Capacitor C3 operates as a low-pass output filter. Capacitor C4 functions as a low-pass loop filter. The DC com-

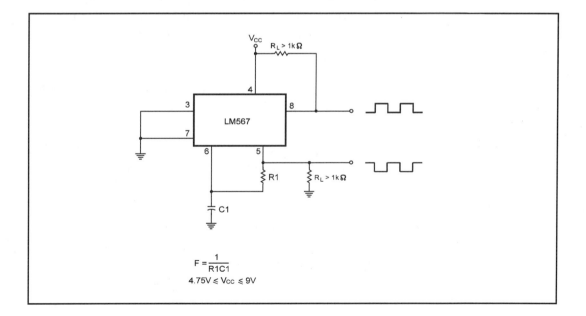

Figure 25-2. Schematic of a quadrature oscillator.

IC Design Projects

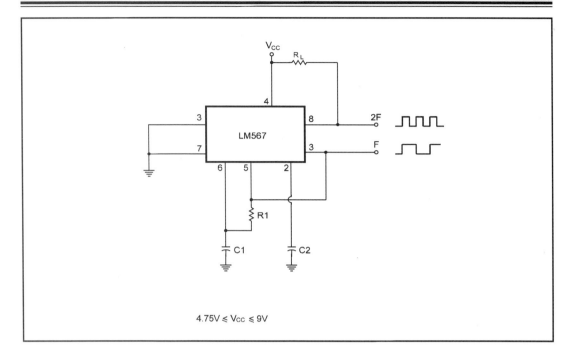

Figure 25-3. Schematic of a two-frequency oscillator.

ponent of the input signal is removed by coupling capacitor C2. The timing resistor, R1, is connected between pin 5 and pin 6. The timing capacitor, C1, is connected between pin 6 and ground. The output is pulled-up to the positive rail of the power supply by pull-up resistor R2.

The center frequency of the tone decoder is equal to the free-running frequency of the VCO. The center frequency, F, is F = 1/R1C1.

The bandwidth of the filter, BW, is obtained from the approximation BW = 1070 x SQRT(V1/FC4), where V1 is the input RMS voltage and where C4 is the capacitance (in uF) at pin 2.

The value in microfarads of the low-pass loop filter capacitor (capacitor connected to pin 2) should be n/F, where *F* is the cutoff frequency of the low-pass loop filter and where *n* ranges between 1300 (for up to 14 percent frequency detection bandwidth) to 62,000 (for up to 2 percent frequency detection bandwidth). The listing of a BASIC program for designing an LM567 tone decoder is given in *Table 25-1*. The program is easy to use. Simply select the appropriate design mode by selecting "1," "2" or "3."

LM567 Tone Decoder IC

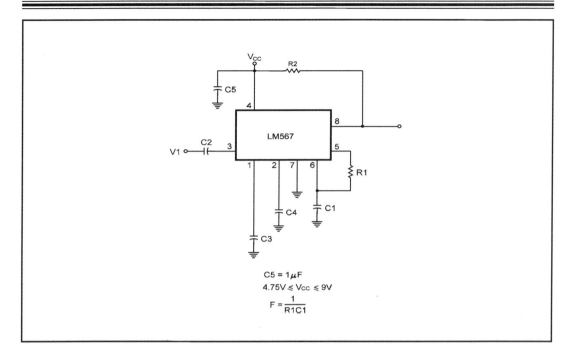

Figure 25-4. Schematic of an LM567 tone decoder.

Exit the program by typing "0." Type "e" when prompted to do so by the onscreen display.

Problems

Problem 25-1. What determines the center frequency of the LM567 tone decoder circuit?

Problem 25-2. What is the frequency range of operation of the LM567 tone decoder IC?

Problem 25-3. Define a quadrature oscillator.

Problem 25-4. What is the frequency of oscillation of an LM567 quadrature oscillator if R1 = 10k and C1 = 0.001 uF?

Problem 25-5. What is the bandwidth of an LM567 tone decoder filter if C4 = 0.01 uF, F = 100 kHz and V1 = 1 volt (RMS)?

Problem 25-6. What value capacitor should be connected to pin 2 of an LM567 tone decoder for a 14 percent frequency detection bandwidth if F = 10 kHz?

IC Design Projects

```
10      REM LM567 TONE DECODER DESIGN PROGRAM
20      FOR I = 1 TO 24:PRINT:NEXT I
30      PRINT "LM567 TONE DECODER DESIGN PROGRAM"
40      PRINT:PRINT:PRINT:PRINT
50      INPUT "DESIGN/EXIT (d/e)?", IN$
60      IF IN$ = "e" THEN END
70      IF IN$ = "d" THEN GOTO 90
80      GOTO 50
90      REM DO DESIGN CALCULATIONS
100     PRINT
110     PRINT "0. END"
120     PRINT "1. FIND R1, GIVEN C1 AND f AS USER ENTRIES"
130     PRINT "2. FIND C1, GIVEN R1 AND f AS USER ENTRIES"
140     PRINT "3. FIND f, GIVEN R1 AND C1 AS USER ENTRIES"
150     PRINT
160     INPUT "CHOOSE:", IN
170     PRINT
180     IF 1N = 0 THEN GOTO 50 ELSE ON IN GOSUB 220, 280, 340
190     GOTO 100
200     REM
210     REM
220     REM FIND R1, GIVEN C1 AND f
230     IPUT "C = ", C1
240     INPUT "f = ", F
250     R1 = 1000000/(F * C1)
260     GOSUB 400
270     RETURN
280     REM FIND C1, GIVEN R1 AND f
290     INPUT "R = ", R1
300     INPUT "f = ", F
310     C1 = 1000000/(F * R1)
320     GOSUB 400
330     RETURN
340     REM FIND f, GIVEN R1 AND C1
350     INPUT "R = ", R1
360     INPUT "C = ", C1
370     F = 1000000/(R1 * C1)
380     GOSUB 400
390     RETURN
400     REM PRINT VARIABLES
410     PRINT
420     PRINT "C1 = "; C1; "uF"; TAB(15); "R = "; R1; "OHMS"; TAB(30)
430     PRINT "f = "; F; "HERTZ"; TAB(45);
440     PRINT
450     RETURN
460     END
```

Table 25-1. Listing of a tone decoder Basic design program.

Chapter 26
◆ Two-Channel Infrared ◆
◆ Remote Control ◆

It is often necessary to switch a device on or off from a distance. This is especially useful if the device is remotely located, such as a satellite in space. For these applications, a remote control system is required. A remote control system consists of a transmitter and a receiver. It is often necessary, for security reasons, to encode the signal at the transmitter then decode the signal at the receiver. The transmitter generates a signal which is then sent into space. The receiver (in this case on the satellite) receives the signal and converts it into an appropriate command, such as turning a device on or off.

This remote control project is a two-channel system. Two devices may be turned on or off using one transmitter and one receiver.

Transmitter Circuit Operation

The transmitter consists of an LM555 timer IC configured as an astable multivibrator, a Darlington pair transistor amplifier, and two infrared LEDs. The transmitter is a portable unit and is therefore powered by a 9-volt alkaline battery, because the alkaline

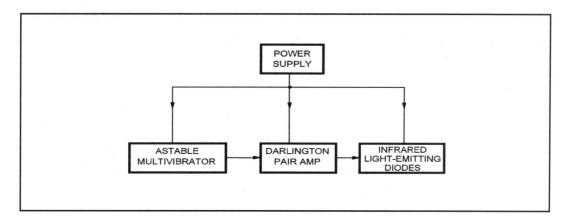

Figure 26-1. Block diagram of the transmitter.

IC Design Projects

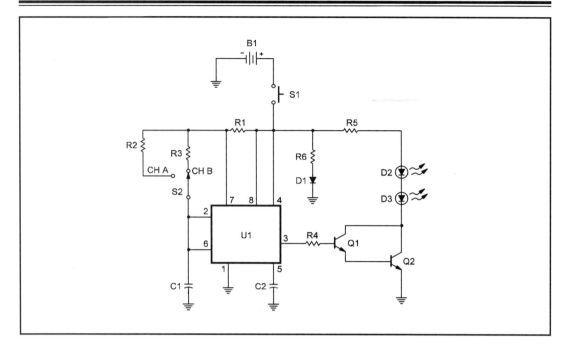

Figure 26-2. Schematic of the transmitter.

battery has an excellent, heavy discharge and rest recovery cycle. *Figure 26-1* is a block diagram of the transmitter. The schematic of the transmitter is shown in *Figure 26-2*. The parts list is shown in *Table 26-1*.

The transmitter must be small enough to fit in a shirt pocket. Therefore, U1, the LM555 timer IC, is configured as an astable multivibrator. The LM555 timer requires only two timing resistors and one time capacitor, C1. Timing resistor R1, and either timing resistor R2 or R3 (as selected by switch S2), determine the frequency of operation of U1. If R2 is selected, U1 operates at 20 kilohertz. If R3 is selected, U1 operates at 10 kilohertz.

The driver amplifier is a Darlington pair transistor amplifier selected for its high input impedance, its high current gain, and its low output impedance. Transistors Q1 and Q2 form the Darlington pair amplifier. Resistor R4 limits the current flow into the base of Q1. The Darlington pair amplifier amplifies the low current output of the timer to a high-enough current to drive the infrared LEDs. Resistor R5 limits the current flow through the infrared LEDs, D2 and D3.

Two-Channel Infrared Remote Control

```
All resistors are ¼ watt @ 5% unless otherwise noted.
All capacitors are rated at 16 volts.

R1, R4, R6:     1k
R2:             3.3k
R3:             6.8k
R5:             51 ohms at 2W
C1, C2:         0.01 uF
Q1:             2N2222A
Q2:             2N3053
D1:             Light-emitting diode
D2, D3:         TIL906-1 or equivalent
U1:             LM555
S1:             N.O. SPST push-button
S2:             SPDT
B1:             9V battery
```

Table 26-1. Parts list for the transmitter.

TIL906-1 infrared LEDs D2 and D3 were chosen because they have a high relative output power. This is very important if a long range is required for the remote control system.

An optional power-on indicator may be included with the transmitter. It consists of LED D1 and a current limiting resistor, R6.

Receiver Circuit Operation

The receiver consists of an infrared phototransistor and a high-gain low-noise amplifier. Each channel of the receiver also contains two decoding circuits, a power drive switch and a relay output. *Figure 26-3* is a block diagram of the receiver. The schematic of the receiver is shown in *Figure 26-4*. The receiver requires a 5-volt power supply. The relay requires a 12-volt power supply.

The infrared phototransistor, Q1, detects the infrared signals of the transmitter. The signal output of the phototransistor is very small, a millivolt or less.

The high-gain amplifier amplifies the small signal to a level of 5 volts, enough to be detected by the first decoders. The high-gain amplifier consists of three common-emitter transistor amplifiers in cascade. Transistors Q2 - Q4, and associated compo-

IC Design Projects

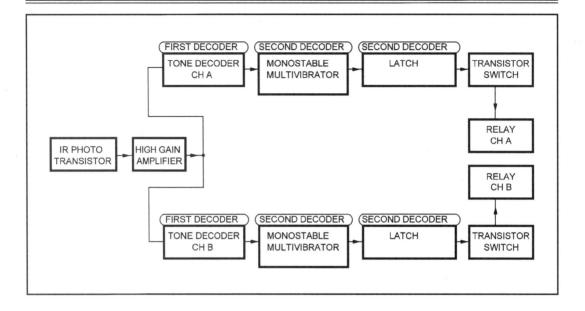

Figure 26-3. Block diagram of the receiver.

nents, form the high-gain amplifier. The common-emitter stage has a relatively high input impedance and a high voltage gain. An operational amplifier was not used because of its gain-bandwidth product. The tone decoder stage can operate at 500 kHz. At this frequency, most operational amplifiers would provide a gain of less than 10. The transistor common-emitter amplifier has a constant high gain up to at least 100 MHz.

The first decoders, U1 and U4, are tone-decoder LM567 ICs. It is tuned to the same frequency of operation as the corresponding transmitter channel by adjusting trimmer potentiometers R14 and R19. The output of the tone decoder is high when no signal or a signal of an incorrect frequency is detected. The LM567 timing components are capacitor C8 and resistors R14 and R15 for channel A, and capacitor C14 and resistors R19 and R20 for channel B.

The outputs of the first decoders are conditioned by U2 and U5, which are LM555 timer ICs configured as monostable multivibrators. The monostable multivibrators also form part of the second decoders by determining the time interval at which the transmitter should be activated. The time interval is supposedly known only to the authorized user of the system.

Two-Channel Infrared Remote Control

The second decoders require that the user activates the transmitter for the correct number of times and at the proper intervals. The second decoders also provide the latching functions required to turn an appliance (for example) on or off. The latching functions are provided by ICs U3 and U6, which are CMOS CD4027 dual J-K master-slave flip-flops.

The power drive switches are inverter-switches, Q5 and Q6, which are capable of driving a load. A CMOS IC cannot drive a load requiring more than one or two milliamperes of current. A transient suppressor should be used to prevent transients in the AC line from "false triggering" the logic circuits. Resistors R18 and R23 limit the current

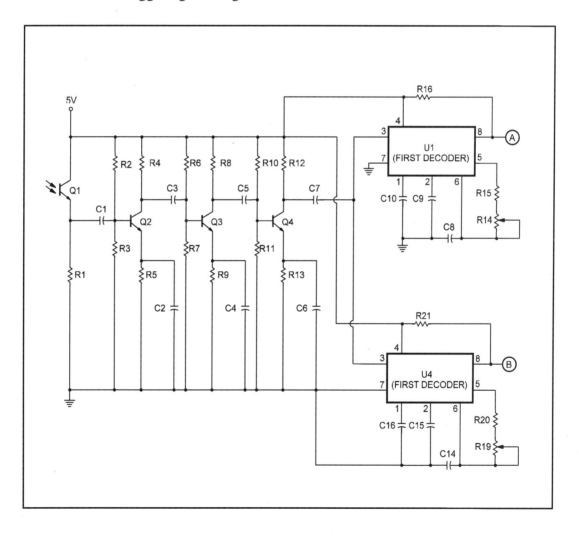

Figure 26-4a. Schematic of the receiver.

IC Design Projects

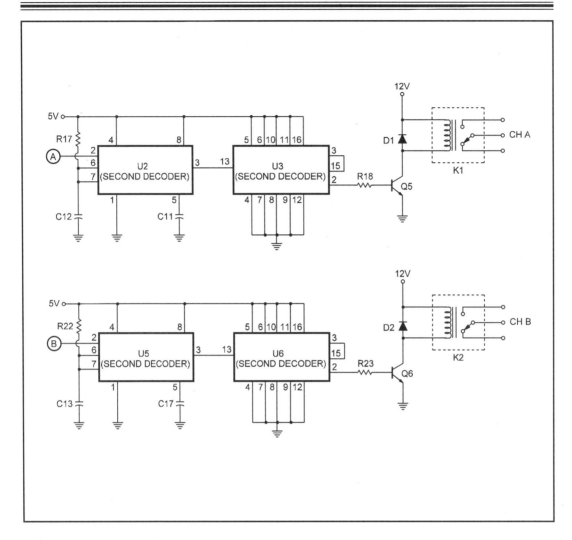

Figure 26-4b. Schematic of the receiver (continued).

flow into the base of transistors Q5 and Q6, respectively. Diodes D1 and D2 protect transistors Q5 and Q6, respectively, from inductive spike voltages generated by relays K1 and K2 at switch-on.

The schematic of the power supply for the receiver is shown in *Figure 26-5*. The transformer T1 steps the household AC line voltage to a low AC voltage. Diodes D1 and D2 rectify the low AC voltage into a pulsating DC voltage. Capacitor C1 reduces the ripple component of the DC voltage. Voltage regulator U1 maintains a constant 5-volt DC output for the receiver circuits. The relay is powered by the 12 volts present at

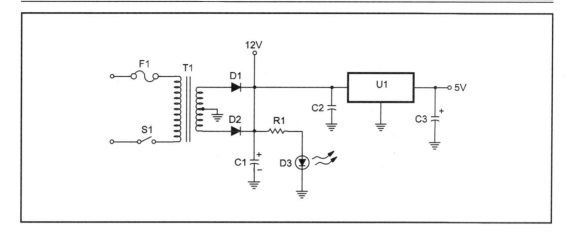

Figure 26-5. Schematic of the power supply for the receiver.

the positive terminal of capacitor C1. Capacitor C3 improves the transient response of voltage regulator U1. LED D3 is a power-on indicator. Resistor R1 limits the current flow through the LED, D3.

Construction

The transmitter and receiver should each be wired on a separate piece of perfboard. Take care to orient all diodes, capacitors and transistor properly. The ICs must not be installed backward or severe damage will result; IC sockets should be used. *Table 26-2* is the parts list for the receiver, and *Table 26-3* is the parts list for the power supply for the receiver.

Care should be taken when soldering because most problems are caused by cold solder joints or short circuits created by solder bridges. If you are inexperienced with soldering, you should practice on a scrap piece of perfboard.

The relay selected can require any coil voltage. If the coil voltage is not 5 volts, then it must be powered by a suitable power supply.

Setup and Use

With the values chosen, the transmitter operates at a frequency of 20 kHz (channel A), and at a frequency of 10 kHz (channel B). The receiver is tuned to the transmitter

IC Design Projects

All resistors are 1/4W @ 5% unless otherwise noted.
All capacitors are rated at 25 volts.

R1:	120 ohms
R2, R6, R10:	75k
R3, R7, R11:	24k
R4, R8, R12:	1.2k
R5, R9, R13:	100 ohms
R14, R19:	5k trimmer potentiometer
R15, R20:	7.5k
R16, R21:	1k
R17, R22:	220k
R18, R23:	10k
C1-C6:	10 uF
C7:	0.47 uF
C8:	0.0047 uF
C9, C15:	1.0 uF
C10, C16:	2.2 uF
C11-C14, C17:	0.01 uF
Q1:	Infrared photo transmitter
Q2-Q4:	2N2222A
Q5, Q6:	2N3053
U1, U4:	LM567
U2, U5:	LM555
U3, U6:	CD4027
D1, D2:	1N4148
K1, K2:	12V relay

Table 26-2. Parts list for the receiver.

frequencies by potentiometers R14 and R19. When properly tuned, the remote control system has a range of at least 15 feet.

This remote control system may be used to turn two appliances on or off. With some minor modifications to channel A of the receiver, the system may also be used as an infrared burglar alarm system for the home or office.

An inverting stage is inserted between ICs U1 pin 8, and U2 pin 2. The output of the monostable multivibrator, U2 pin 3, is then connected to R18, thus eliminating the latch, U3, which is part of the second decoder.

The system is turned on when the personnel leave the office, or when the home is left unattended. As long as the receiver detects the transmitter signal, there is no power applied to the load, which can be a siren, an automatic telephone dialer, a relay, etc.

When a burglar breaks the invisible infrared beam between the transmitter and receiver, power is applied to the alarm. The system will also automatically reset or turn off the alarm after a predetermined time interval. The remote control should provide years of reliable service.

All resistors are 1/4W @ 5% unless otherwise noted.
All capacitors are rated at 25 volts.

R1:	1.5k
C1:	1000 uF
C2:	0.33 uF tantalum
C3:	1.0 uF tantalum
U1:	LM7805
D1, D2:	1N4001
D3:	Light-emitting diode
T1:	18 V.C.T. secondary at ½ ampere
F1:	½ ampere slow blow
S1:	SPST

Table 26-3. Parts list for the power supply for the receiver.

Appendix
♦ Problem Solutions ♦

Chapter 1

Problem 1-1. Electronic circuits require power supplies because they need a stable DC source for a stable "Q" or operating point.

Problem 1-2. Batteries are not the power supplies of choice because they are expensive. They have a limited life span and most batteries cannot be recharged.

Problem 1-3. The three parts of an AC-to-DC power supply are the transformer, the rectifier and the regulator.

Problem 1-4. Power supplies that operate at 400 Hz are smaller and lighter than those operating at 60 Hz.

Problem 1-5. When a voltage is applied to the primary windings of a transformer, there is a current flow through the primary windings. The primary current of a transformer induces a magnetic flux flow through the magnetic core. The magnetic flux induces a current flow through the secondary windings of the transformer. The secondary current generates a voltage across the secondary windings of the transformer.

Problem 1-6. P = IV, substituting values 50 = 25I, from which I = 2 amperes.

Problem 1-7. A step-up transformer has a secondary voltage that is greater than its primary voltage. A step-down transformer has a secondary voltage that is less than its primary voltage.

Problem 1-8. An isolation transformer has equal primary and secondary voltages.

Problem 1-9. Regulators are used in power supplies because electronic circuits require power supplies with ripple factors of much less than 48.2 percent.

Problem 1-10. The regulator of an AC-to-DC power supply is essentially a low-pass filter with a cutoff frequency of much less than 60 Hz.

Problem 1-11. $CdV = I_L dt$.
By substitution, C = dt = 0.004167 F = 4167 uF.

Problem 1-12. Let I_z = 5 mA:
R1 = (V1 - V2)/(I_L + Iz) = (9 - 5)/(0.050 + 0.005) = 4/0.055 = 73 ohms.

IC Design Projects

$P_{R1} = (V1 - V2)(I_L + I_z) = (9 - 5)(0.050 + 0.005) = 4 \times 0.055$
$= 220$ mW.
$P_{D1}(max) = V2(I_L + I_z) = 5(0.050 + 0.005) = 5 \times 0.055 = 275$ mW.

Problem 1-13. $V3 = V2 - V_{BE}$, from which $V2 = V3 + V_{BE} = 5 + 0.6 = 5.6$ volts.
Therefore, a 5.6-volt zener diode is required;
$I_B = I_L/B = 1/50 = 20$ mA.
$R1 = (V1 - V2)/(I_z + I_B) = (9 - 5.6)/(0.005 + 0.020) = 3.4/0.025$
$= 136$ ohms.
$P_{R1} = (V1 - V2)(I_z + I_B) = (9 - 5.6)(0.005 + 0.020) = 3.4 \times 0.025$
$= 85$ mW.
$P_{D1}(max) = V_z I_z = 5.6 \times 0.005 = 28$ mW.

Chapter 2

Problem 2-1. The voltage regulator reduces the AC component of an AC-to-DC power supply output DC voltage.

Problem 2-2. The two types of regulators are the switching regulator and linear regulator.

Problem 2-3. The switch-mode power supply is small in size and has a high conversion efficiency. They are used in commercial applications, such as in computers.

Problem 2-4. The linear power supply is easy to design and build. It is the choice of hobbyists.

Problem 2-5. The main building block of an IC linear regulator is the DC operational amplifier.

Problem 2-6. An operational amplifier compares the reference voltage with a fraction of the output voltage. The operational amplifier controls a series-pass element to regulate the output voltage, depending on the differential voltage between its input terminals.

Problem 2-7. External series-pass transistors make it possible to design power supplies that can handle higher currents than the IC regulator specified current.

Problem 2-8. Switching regulators are useful in battery-operated equipment, space vehicles, commercial and industrial applications.

Problem 2-9. Vout = Vin(t_{ON}/T) = 20(0.00001/0.00002) = 20 × 0.5 = 10 volts.

Problem 2-10. The LM723 regulator consists of a voltage reference amplifier, a temperature-compensated reference-voltage zener diode, a second

Problem Solutions

zener diode, an error amplifier, a current-limiter transistor, and a series-pass transistor.

Problem 2-11. $R_{SC} = V_{BE}/I_{OUT}(MAX) = 0.65/1 = 0.65$ ohms.

Problem 2-12. The minimum output voltage is two volts if pin 7 of an LM723 regulator is connected to ground.

Problem 2-13. Pin 7 of an LM723 regulator should be connected to a -3.6 volt source, for a minimum output voltage of zero volts.

Problem 2-14. The second zener diode of an LM723 regulator is usually used to offset the output voltage, when the LM723 regulator is configured as a negative voltage regulator.

Problem 2-15. Vout = Vref(1 + [R2/R1]) = 1.25(1 + [1000/220]) = 1.25(1 + 4.55) = 1.25 x 5.55 = 6.93 volts

Problem 2-16. R2 = R1([Vout/Vref] - 1) = 120([-12/-1.25]-1) = 120(9.6-1) = 120 x 8.6 = 1032 ohms

Chapter 7

Problem 7-1. TTL stands for transistor-transistor logic.

Problem 7-2. A common-emitter circuit can quickly discharge a capacitive load.

Problem 7-3. An emitter-follower circuit can quickly charge a capacitive load.

Problem 7-4. In steady state, Q3 can sink current from a load.

Problem 7-5. In steady state, Q4 can source current to a load.

Problem 7-6. Diode D1 prevents Q4 from saturating when the input voltage is high.

Problem 7-7. A TTL inverter is converted into a two-input NAND gate by adding a second base-emitter junction to Q1.

Problem 7-8. A power supply of 5V +/-0.5V is required for 7400 series TTL ICs.

Problem 7-9. Propagation delay is the time between the 1.5-volt point of corresponding edges of the input and output waveforms.

Problem 7-10. The zero-level and one-level noise margins are both 0.4 volts.

Problem 7-11. The propagation delay of a typical TTL gate is 18 nanoseconds.

Problem 7-12. The fan-in of a TTL gate is 8.

Problem 7-13. The fan-out of a TTL gate is 12.

Problem 7-14. A TTL gate can dissipate five milliwatts of power when the output voltage is high and 16.7 milliwatts when the output voltage is low.

Problem 7-15. The switching speed of a TTL gate can be improved by reducing the resistor values of the gate, and by preventing Q1, Q2 and Q3 from saturating.

IC Design Projects

Problem 7-16. High-speed TTL gates cannot dissipate as much power as regular TTL gates because the resistor values are reduced.

Problem 7-17. A Schottky transistor cannot saturate because a Schottky diode is placed across its base-collector junction. The diode clamps the voltage across the base-collector junction to 0.4 volts, which is less than the 0.6 volts required for transistor saturation.

Chapter 8

Problem 8-1. CMOS stands for complementary metal oxide silicon.

Problem 8-2. CMOS ICs consume little power, have excellent noise immunity, and can operate from 3 to 18 volts.

Problem 8-3. A MOS transistor is fabricated by superimposing several layers of conducting, insulating and transistor-forming materials.

Problem 8-4. When an input voltage is applied to the gate of a MOS transistor, current flows through a conducting channel from the source to the drain of the MOS transistor.

Problem 8-5. The gate of a MOS transistor is an area separating the drain and the source. The gate is a conducting electrode.

Problem 8-6. The gate-to-substrate potential difference controls the current flow from the source to the drain of a MOS transistor.

Problem 8-7. A MOSFET should be biased such that the drain-to-substrate and source-to-substrate junctions are reverse-biased. In a P-channel device, the substrate is always the most positive voltage. In an N-channel device, the substrate is always the most negative voltage.

Problem 8-8. The inverter switches faster, dissipates more power, and has greater noise immunity when the power supply voltage is increased within the 3- to 18-volt range.

Problem 8-9. If the N-channel and P-channel devices are matched, the switching threshold voltage is one-half the power supply voltage.

Problem 8-10. Manufacturers specify minimum and maximum transfer characteristics because the N-channel and P-channel devices are not matched in practical CMOS ICs.

Problem 8-11. CMOS circuits operate from 3 volts DC to 18 volts DC.

Problem 8-12. Noise margin is the allowable noise voltage on the input of a gate so that the output of the gate is not affected.

Problem 8-13. CMOS noise margin specifications are:
$V_{DD} = 5V$, $0NM = 1NM = 0.5V$.

Problem Solutions

$V_{DD} = 10V$, 0NM = 1NM = 1.0V.
$V_{DD} = 15V$, 0NM = 1NM = 1.0V.

Problem 8-14. A CMOS gate dissipates several nanowatts when it is not in its switching mode. A CMOS gate dissipates several milliwatts when it is in its switching mode.

Problem 8-15. Some precautions when handling CMOS ICs are:
- A. Connect all unused input leads either to the positive power supply rail or to the circuit ground.
- B. Never connect output terminals together to create wired-logic circuits.
- C. Store unused CMOS devices in conductive foam.
- D. The soldering iron should have a grounded tip.
- E. Touch the circuit ground with one hand and pick up the CMOS device with the other hand.

Chapter 11

Problem 11-1. The operational amplifier (op-amp) got its name because it was originally used to perform mathematical operations on electrical signals.

Problem 11-2. An op-amp has a differential amplifier input stage, a level-shifter stage, and a power amplifier stage.

Problem 11-3. The stages of an op-amp are direct coupled. There are NO coupling capacitors used; therefore, op-amps can operate down to DC or zero frequency.

Problem 11-4. The ideal op-amp has infinite input impedance, infinite gain, and zero output impedance.

Problem 11-5. Open-loop gain is gain without a feedback network.

Problem 11-6. When the two input voltages of an op-amp are equal, the output voltage is zero volts.

Problem 11-7. A practical op-amp has an input impedance of at least one megohm, an open-loop gain of about 200,000, and an output impedance of about 100 ohms.

Problem 11-8. Most op-amps require a dual or bipolar power supply.

Problem 11-9. Offset voltage is the output voltage generated by the op-amp when the input terminals are grounded.

Problem 11-10. Input offset voltage is the differential input voltage that must be applied to the input terminals of an op-amp for a zero output voltage.

IC Design Projects

Problem 11-11. Slew rate is the maximum rate of change of the op-amp output voltage in response to a square wave differential-mode input signal.

Problem 11-12. CMRR is the ability of an op-amp to cancel out common signals that are fed to its two input terminals.

Problem 11-13. GBW = 1,000,000 and F = GBW/GAIN = 1,000,000/10 = 100,000 Hz.

Problem 11-14. An op-amp can be destroyed if:
    A. An excessive power supply voltage is used.
    B. An excessive input voltage.
    C. The connections to the power supply are reversed.
    D. The load of the op-amp is short-circuited.

Problem 11-15. An op-amp operates in its linear mode when its output voltage is directly proportional to its input voltage.

Problem 11-16. An op-amp operates in its nonlinear mode when its output is in saturation.

Problem 11-17. $V_{OUT} = -R2(V_{IN})/R1 = -10,000 \times 0.1/1000 = -1$ volt and $R_{IN} = R1 = 1$ kohm.

Problem 11-18. $V_{OUT} = V_{IN}(1 + (R2/R1)) = 0.1(1 + (10,000/1000)) = 0.1 \times 11 = 1.1$ volts.
$R_{IN} = AR1/(1 + (R2/R1)) = 100,000 \times 1000/(1 + (10,000/1000)) = 9,090,909$ ohms.

Problem 11-19. $V_{OUT} = -V_{IN1}(R4/R1) - V_{IN2}(R4/R2) - V_{IN3}(R4/R3)$
$= -V_{IN1} - V_{IN2} - V_{IN3}$.

Problem 11-20. $V_{OUT} = R4(R1 + R2)V_{IN2}/R1(R3 + R4) - R2V_{IN1}/R1$
$= R4(V_{IN2} - V_{IN1})/R1$.

Problem 11-21. In mathematics, the integral of cosine is sine. The output of an op-amp integrator is the integral inverted. Therefore, the output voltage is an inverted sine wave.

Problem 11-22. $T = R1C1 = 1,000 \times 0.01$ uF $= 10$ usec.

Problem 11-23. A low-pass filter passes only those frequencies below its cutoff frequency. A high-pass filter passes only those frequencies above its cutoff frequency. A bandpass filter passes only those frequencies between its two cutoff frequencies. A notch filter passes all frequencies EXCEPT those between its two cutoff frequencies.

Problem 11-24. The output voltage of an op-amp comparator is equal to the positive supply voltage, when the noninverting input voltage is greater than the reference voltage on its inverting input.

Problem 11-25. $T = -R4C2V_{REF}/V_{FEEDBACK} = -1,000 \times 0.01$ uF $\times 10/1 = -0.1$ msec. The minus sign indicates an inverted output.

Problem Solutions

Chapter 15

Problem 15-1. PLL means phase-locked loop.

Problem 15-2. PLL is used to synchronize one sinusoidal waveform with another.

Problem 15-3. The components of a PLL are:
- A. VCO; generates a reference signal.
- B. Phase comparator; compares reference and input signals; generates an error voltage based on the difference between reference and input signals.
- C. Loop filter; smooths error voltage; error voltage is used to control VCO.

Problem 15-4. The error voltage causes the VCO to change its frequency until it equals the frequency of the input signal.

Problem 15-5. The lock range is the range of frequencies over which the PLL can track changes in the input frequency.

Problem 15-6. The PLL is not operating linearly when it is not locked onto the input signal.

Problem 15-7. $F = Nf = 10 \times 10$ MHz $= 100$ MHz

Problem 15-8. $F = 0.3/R1C1 = 0.3/0.000001 = 300$ kHz.

Problem 15-9. The VCO of an LM565 cannot operate separately because its control voltage input is connected to the amplifier output through an internal 3.6 kilohm resistor.

Problem 15-10. No transformers or tuned circuits are required in an FM detector circuit using an LM565 PLL IC.

Problem 15-11. Capture range is the range of frequencies over which the PLL can lock on an input signal, which initially is out of lock.

Problem 15-12. The difference between the two phase comparators, or a CD4046 PLL IC, are:
- A. Phase comparator I is immune to noise; phase comparator II is sensitive to noise.
- B. Phase comparator I can lock onto the harmonics of the VCO frequency; phase comparator II cannot lock onto the harmonics of the VCO frequency.
- C. Phase comparator I requires a square wave input with a 50 percent duty cycle; phase comparator II accepts input signals of any duty cycle.

Problem 15-13. F1 = 1/(R1(C1 + 32 pF)) = 1/0.00001 = 100 kHz.
Problem 15-14. F2 = 1/(R2(C1 + 32 pF))) = 1/0.000001 = 1.0 MHz.
Problem 15-15. F = F1 + F2 = 1.1 MHz.

Chapter 19

Problem 19-1. The LM555 timer IC can operate in either the astable mode or the monostable mode.
Problem 19-2. The LM555 timer is used in timing circuits, pulse-generating circuits, pulse detector circuits, pulse-width modulator circuits, and pulse-position modulator circuits.
Problem 19-3. The LM555 timer is classified as a linear IC because it can be triggered by either a linear signal or a digital signal.
Problem 19-4. No, because the output is always a digital signal.
Problem 19-5. A comparator is a high-gain differential amplifier.
Problem 19-6. In an LM555, the threshold is the input to the upper comparator, and the trigger is the input to the lower comparator.
Problem 19-7. A comparator changes its output state when its input signal equals its reference voltage.
Problem 19-8. Transistor Q1 discharges the external timing capacitor.
Problem 19-9. The control flip-flop drives transistor Q1.
Problem 19-10. Transistor Q2 accepts an input pulse and resets the control flip-flop.
Problem 19-11. An astable multivibrator is a free-running oscillator.
Problem 19-12. F = 716 Hz. DC = 50.25%.
Problem 19-13. R1 > 75 ohms.
Problem 19-14. A monostable multivibrator has one stage state: the off state. When it is triggered by an input pulse, it enters its unstable state for a time interval determined by the external timing components. The monostable multivibrator then returns to its stable state.
Problem 19-15. T = 517 usec.
Problem 19-16. False triggering is avoided in LM555 timer circuits by connecting the reset terminal (pin 4) to the Vcc terminal (pin 8) of the LM555 timer IC.
Problem 19-17. The monostable is converted to a pulse-width modulator by applying a continuous pulse train to pin 5 (control voltage terminal) of the LM555 timer.
Problem 19-18. The astable is converted to a pulse-position modulator by applying a signal to pin 5 (control voltage terminal) of the LM555 timer.

Problem Solutions

Problem 19-19. The monostable is converted to linear ramp generator by replacing pull-up resistor R1 (also serves as the timing resistor) with a constant current source.
Problem 19-20. T = 58 usec.
Problem 19-21. R2 is less than R1/2.

Chapter 25

Problem 25-1. The VCO and the external timing components determine the center frequency of the LM567 tone decoder circuit.
Problem 25-2. The LM567 tone decoder IC can operate in the frequency range of 0.01 Hertz to 500 kilohertz.
Problem 25-3. A quadrature oscillator has two outputs that are 90 degrees out of phase with each other.
Problem 25-4. F = 1/R1C1 = 100 kHz.
Problem 25-5. BW = 1070 x SQRT(V1/FC4) = 34 hertz.
Problem 25-6. C = n/F = 0.13 uF.

◆ Glossary ◆

Amplifier: A device that provides a gain of more than unity.

Astable Multivibrator: A two-stage oscillator circuit that continuously switches between its two states. It is also called a free-running oscillator.

Bistable Multivibrator: A two-stage oscillator circuit where the output is fed back to the input. It has two stable states. It is also called a flip-flop.

Close-Loop Gain: The gain of an amplifier with a feedback loop.

Common-Mode Gain: The ratio of the output voltage of a differential amplifier to the common-mode input voltage. The common-mode gain of an ideal differential amplifier is zero.

Common-Mode Input Voltage: An input voltage common to the two inputs of a differential amplifier.

Common-Mode Rejection Ratio: The ratio of differential-mode gain to common-mode gain.

Compensation: The shaping of the op-amp frequency response in order to achieve stable operation in a particular circuit.

Current Clamping: The output-current limiting feature of some op-amps.

Cycle: One complete excursion of the instantaneous value of the induced EMF. The portion of a periodic waveform between two adjacent corresponding points of the waveform.

Differential Amplifier: An amplifier that amplifies the voltage difference between its two inputs.

Differential-Mode Gain: The ratio of the output voltage of a differential amplifier to the differential-mode input voltage.

Differential-Mode Input: The voltage difference between the two inputs of a differential amplifier.

Duty Cycle: The amount of time during which a device or signal is active.

Flip-Flop: A bistable multivibrator. A one-bit memory.

Frequency: The number of cycles completed in one second. The reciprocal of a period.

Inductive Kickback: Large counter EMF generated across an inductance when an existing current is interrupted.

Input Bias Current: The current that must be applied to each input of an IC op-amp to assure proper biasing of the differential input stage transistors.

Input Offset Current: The difference between the input bias currents flowing into each input of an IC op-amp, when the output is at zero volts.

Input Offset Voltage: The voltage that must be applied across the two inputs of an op-amp in order to produce zero volts at the output.

Inverting Amplifier: An amplifier whose output is 180 degrees out of phase with its input.

Monostable Multivibrator: A two-stage oscillator that provides positive feedback. It has only one stable state.

Noninverting Amplifier: An amplifier whose output signal is in phase with its input signal.

Open-Loop Gain: The gain of an amplifier without a feedback loop.

Operational Amplifier (Op-Amp): An IC differential amplifier with high gain, high input impedance, and low output impedance.

Oscillator: A circuit that generates a periodic waveform.

Glossary

Output Offset Voltage: The output voltage of a negative feedback op-amp when the input voltage is zero.

Period: The time taken to complete one cycle of a waveform. The reciprocal of frequency.

Periodic: A waveform that repeats itself.

Phase-Locked Loop (PLL): A circuit used to connect one sinusoidal waveform to another sinusoidal waveform.

Potentiometer: A variable resistor.

Primary: Input winding of a transformer.

Pulse-Position Modulation: The position of the output pulse depends on the amplitude of the input signal.

Pulse-Width Modulation: The width of the output pulse is proportional to the amplitude of the input pulse.

Quadrature Oscillator: An oscillator that has two outputs that are 90 degrees out of phase with each other.

Rectifier: A device that conducts current in one direction and blocks the flow of current in the other direction.

Rheostat: A variable resistor.

Secondary: Output winding of a transformer.

Sinusoidal: Waveform with the shape of a sine wave.

Slew Rate: The maximum rate of change of the output voltage of an op-amp as it swings from positive to negative, or from negative to positive, in response to a square wave differential-mode input.

Stepdown Transformer: Transformer with fewer turns on the secondary than on the primary.

Step-Up Transformer: Transformer with more turns on the secondary than on the primary.

Time Constant: The time required for the transient current and voltages in a series RC circuit to reach 63.2 percent of their final values; $T = RC$

◆ Bibliography ◆

The ARRL Handbook, 70th Edition, The American Radio Relay League, Connecticut, 1993.

COS/MOS Integrated Circuits, RCA Corporation, New Jersey, 1974.

Dorf, R.C. *Modern Control Systems*, Third Edition, Addison-Wesley Publishing Company, Massachusetts, 1983.

Lenk, J.D. *Manual For Integrated Circuit Users*, Reston Publishing Company, Inc., Virginia, 1973.

Linear Integrated Circuits, Motorola Inc., Arizona, 1979.

Linear Integrated Circuits, National Semiconductor Corporation, California, 1973.

Melen, R. and Garland, H. *Understanding IC Operational Amplifiers*, Howard W. Sams & Co., Inc., Indianapolis, 1973.

Sedra, A.S. and Smith, K.C., *Micro-Electronic Circuits*, Holt, Rinehart and Winston, New York, 1982.

Stremler, F.G., *Introduction To Communication Systems*, Second Edition, Addison-Wesley Publishing Company, Massachusetts, 1982.

The TTL Data Book for Design Engineers, Texas Instruments Incorporated, Texas, 1973.

◆ Index ◆

SYMBOLS

2N3055 TRANSISTOR 183
5-VOLT POWER SUPPLIES 65
5-VOLT POWER SUPPLY 57, 60, 63, 65
5-VOLT SUPPLY 57
7400 73, 78

A

AC 8
AC BRIDGE 91
AC COMPONENT 5, 10, 12, 27, 49, 54
AC CORD GROMMET 199
AC HOUSEHOLD LINE VOLTAGE 54
AC HUM 136
AC LINE 221
AC LINE CORD GROMMET 207
AC LINE VOLTAGE 49, 67, 137, 154, 159, 197, 205, 222
AC OPERATED POWER SUPPLY 201
AC POWER 195, 198
AC POWER SUPPLY 203
AC SOCKET 197, 199, 205-208
AC VOLTAGE 5, 49, 54, 57, 67, 91, 124, 137, 154, 159, 187, 197, 205, 222
AC VOLTMETER 134
AC-COUPLED FLIP-FLOP CIRCUIT 115
AC-OPERATED EQUIPMENT 135
AC-TO-DC POWER SUPPLY 5
ACOUSTICAL OUTPUT 202
ACOUSTICS 119, 137
ACTIVE MODE 74, 77
ADJUSTABLE POWER SUPPLY 36
ADJUSTABLE TERMINALS 44
ADJUSTABLE TIMING DELAY 205
ADJUSTABLE VOLTAGE REGULATORS 32
ADJUSTMENT CURRENT 46, 47
ADJUSTMENT TERMINALS 47
AIRCRAFT 5
ALARM 201, 225
ALARM OUTPUT 157
ALCOHOL 135
ALKALINE BATTERY 217
ALLIGATOR CLIP 102, 163
ALPHA 129
ALPHA BRAIN WAVES 127, 130, 135
ALPHA BRAIN-WAVE FEEDBACK MONITOR 127, 131, 133, 135, 136
ALPHA POSITION 135
ALTERNATING CURRENT (AC) 5, 6, 14
ALTERNATING LINE VOLTAGE 6

ALTERNATING VOLTAGE 6, 7
ALUMINUM 52, 55, 68, 101, 102, 190
ALUMINUM FOIL 89
AM/FM MODULATOR 131
AMBIENT-REFLECTED SOUND 126
AMPERES 12, 45, 63
AMPLIFICATION 110
AMPLIFICATION FACTOR 111
AMPLIFIED BRAIN WAVES 132
AMPLIFIER 109, 121-123, 128, 129, 132, 134, 147, 148, 158, 162, 219, 220, 237
AMPLIFIER OUTPUT 147
AMPLITUDE 122, 153, 156
AMPLITUDE OF OSCILLATION 31
ANALOG BUILDING BLOCK 27
ANALOG CIRCUIT 156
ANALOG CIRCUITS 153, 156
ANALOG COMPUTERS 107
ANALOG DC VOLTMETER 156
ANODE 38
ANTILOCK GATING 195
ANTILOG AMPLIFIERS 116
APPLIANCE 221, 224
APPLIANCE CONTROLS 83
APPLICATIONS 147, 205, 213
ARMATURE VOLTAGE 181
ASTABLE MODE 167
ASTABLE MULTIVIBRATOR 92, 93, 95, 102, 115, 157, 162, 169, 170, 173, 188-192, 195, 201, 217, 218, 237
ASTABLE MULTIVIBRATOR CIRCUIT 95
ASTABLE OPERATION 169
ATTENUATION 31, 139
ATTENUATION FACTOR 189
AUDIBLE LOGIC 161
AUDIBLE LOGIC PROBE 161, 162, 163, 164
AUDIBLE LOGIC PROBE INPUT 163
AUDIBLE LOGIC PROBE TIP 163, 164
AUDIO AMPLIFIER 131, 187, 188, 189, 190
AUDIO AMPLIFIER IC 190
AUDIO APPLICATIONS 123
AUDIO CIRCUITS 111
AUTOCYCLER ASTABLE MULTIVIBRATOR 93
AUTOCYCLING CIRCUIT 92
AUTOMATIC LIGHT TIMER 195, 198, 199, 200
AUTOMATIC LIGHT TIMER CIRCUIT 195
AUTOMATIC TELEPHONE DIALER 224
AUTOMOBILE 157
AUTOMOTIVE ELECTRONICS 83
AUTORANGING CIRCUIT 95
AVAILABILITY 28

B

BACK-UP BATTERY 195, 198
BAND 135
BAND SIGNALS 212
BANDPASS FILTER 115, 129, 131, 132, 136
BANDPASS FILTER SWITCH 135
BANDPASS SWITCH 135
BANDWIDTH 109, 144, 146, 211, 212, 214
BANDWIDTH SETTINGS 135
BASE 16, 36, 74, 75, 162, 218, 222
BASE BIAS CURRENTS 110
BASE CURRENT 16, 202
BASE CURRENT DRIVE 197
BASE CURRENT FLOW 183
BASE DRIVE 59, 77, 80, 197
BASE DRIVE CURRENT 16
BASE SPEED 181
BASE TERMINAL 169
BASE VOLTAGE 52, 76, 79
BASE-COLLECTOR JUNCTION 74, 76, 80
BASE-EMITTER JUNCTION 50, 74, 76-78, 112
BASIC PROGRAM 17, 173, 214
BASIC RECTIFIER CIRCUITS 7
BASS BOOST 122, 137, 139
BASS CONTROL 126, 137
BASS CONTROL CIRCUIT 139
BASS CUT 137, 139
BASS TONE CONTROL CIRCUIT 139
BATTERY 5, 49, 51, 129, 133, 134, 151, 195, 198, 201, 203
BATTERY CHARGER 49, 51, 52
BATTERY CHARGER CABINET 52
BATTERY CHARGING RATE 49
BATTERY CURRENT DRAIN 203
BATTERY POWERED 157
BATTERY VOLTAGE 30
BATTERY-POWERED CIRCUITS 110, 151
BATTERY-POWERED EQUIPMENT 30
BCD 95
BCD OUTPUTS 93
BEEP 135
BELL LABORATORIES 83
BETA 16, 129
BETA BRAIN WAVES 127
BIAS 84
BIAS CURRENT 138
BILATERAL SWITCH 157
BILATERAL SWITCH CONTROL TERMINAL 157
BINARY CIRCUIT 168
BINARY COUNTER 94
BIOFEEDBACK 130
BIOPOTENTIAL SIGNALS 134
BIOPOTENTIAL VOLTAGE SOURCE 128
BIOPOTENTIALS 127, 128, 129
BIPOLAR POWER SUPPLY 20, 53, 54, 55, 109, 110, 124, 146, 147, 154, 155
BIPOLAR TRANSISTOR 83, 109
BISTABLE MULTIVIBRATOR 115, 237
BLACK 102, 163, 164
BLEEDER RESISTORS 59
BLOCK CURRENT 94
BLOCK DIAGRAM 119, 120, 145, 167, 218, 219
BODY SIGNALS 128
BRAIN 127, 129, 134
BRAIN VOLTAGES 129
BRAIN WAVE CHARACTERISTICS 127
BRAIN WAVES 127, 129, 130, 132, 135
BRAIN-WAVE BIOPOTENTIALS 128, 129
BRAIN-WAVE FEEDBACK MONITOR 131, 133
BRAIN-WAVE FREQUENCY 132
BRAIN-WAVE MONITOR 128, 129, 132, 133, 134, 135
BRAIN-WAVE SIGNAL 128, 134
BREADBOARD 164
BRIDGE AMPLIFIER 123
BRIDGE CONFIGURATION 10
BRIDGE POWER AMPLIFIER 124
BRIDGE RECTIFIER 49, 154
BUFFER 113, 198
BUFFER AMPLIFIER 120, 153, 156
BUFFER STAGES 109
BUFFER/DRIVER 196
BUFFERED INPUT STAGES 119
BURGLAR 225
BURST 103
BUSHINGS 60

C

CABINET 101, 102
CABINET DESIGN 190
CALIBRATION 199
CALIBRATION PROCEDURE 95
CAPACITANCE 12, 13, 91, 92, 95, 212, 214
CAPACITANCE MEASUREMENT 92
CAPACITANCE METER 91, 94
CAPACITANCE METER CIRCUITS 91
CAPACITIVE LOAD 74
CAPACITOR 12, 14, 17, 33, 39, 49, 52, 54, 55, 61, 67, 91, 93, 95, 96, 101, 102, 109, 114, 122, 124, 125, 133, 137-139, 144, 146, 147, 148, 150, 153, 155, 157-160, 162, 167, 169, 170, 187, 190, 195, 197, 198, 205, 207, 212, 213, 214, 220, 222, 223
CAPACITOR CONNECTED 212
CAPACITOR FILTER 60
CAPACITOR LEAD 95
CAPACITOR REGULATOR 12, 13
CAPACITOR VALUES 173
CAPACITOR VOLTAGE 22, 52
CAPACITOR WORKING VOLTAGES 55
CAPTURE CHARACTERISTICS 147
CAPTURE PROCESS 143
CAPTURE RANGE 143, 144, 146, 148
CAR BATTERY 201

Index

CARRIER DEVICE 83
CARRIER-CURRENT REMOTE CONTROL 213
CARRY-OUT SIGNAL 196
CASCADE 132, 219
CASCADING 111
CASE 103
CASSETTE DECK 137
CD 139
CD PLAYER 137, 138
CD-QUALITY RECORDINGS 137
CD4013 98
CD4017 195, 196
CD4027 221
CD4046 148, 150, 151, 157, 161
CD4050 196
CDS 137
CENTER FREQUENCY 150, 162, 211, 212, 214
CENTER TAP 94
CENTER VOICE CHANNEL 120
CENTER-TAPPED SECONDARY 154
CENTER-TAPPED SECONDARY WINDING 8
CENTER-TAPPED STEP-DOWN TRANSFORMER 187
CENTER-TAPPED TRANSFORMER 10
CEREBRAL FLUIDS 134
CHANNEL 83, 120, 121, 124, 137, 139, 219
CHANNEL AMPLIFIERS 119, 123
CHANNEL INPUTS 139
CHANNEL SIGNAL 123
CHARGE 12, 88, 133, 144, 173
CHARGE TIME 170
CHARGING RATE 49, 51
CHARGING RATES 52
CHARGING TIME 49
CHART RECORDER 135
CHASSIS-MOUNT FUSE HOLDER 199
CHIRPING SOUND 158, 160
CHOPPER 181
CHOPPER SPEED CONTROL CIRCUIT 181
CHORDS 185
CIRCUIT 8, 10, 12, 20, 31, 38, 42, 53, 54, 57, 61, 63, 67, 68, 75-77, 91, 94, 98, 102, 103, 108, 111, 113, 115, 120, 131, 132, 136-138, 144, 150, 151, 153, 156, 157, 161-163, 167, 169, 173, 176, 177, 181, 184, 187, 195, 199, 201, 203-205
CIRCUIT CURRENT 15
CIRCUIT DESIGNER 144
CIRCUIT GROUND 133
CIRCUIT OPERATION 49, 61
CIRCUIT STABILITY 136
CIRCUIT UNDER TEST 97
CIRCUIT VOLTAGES 75, 76
CIRCUIT-GENERATED NOISE 136
CLOCK 83, 92
CLOCK ENABLE 195
CLOCK INPUTS 98
CLOCK PULSE 98, 195, 196, 199

CLOCK PULSE CYCLE 195
CLOSE-LOOP GAIN 237
CLOSED LOOP SYSTEM 146
CLOSING DOOR 120
CMOS 73, 83, 85, 89, 99, 161, 221
CMOS CIRCUITS 85, 97
CMOS DECADE COUNTERS 195, 196
CMOS DEVICE 87-89
CMOS GATES 87, 88, 89
CMOS HEX BUFFER 196
CMOS IC 163, 221
CMOS ICS 85, 86, 88, 89, 98, 157, 159, 163
CMOS INTEGRATED CIRCUIT 98
CMOS INVERTER 84, 85
CMOS LOGIC 83, 87, 89
CMOS LOGIC CIRCUITS 86
CMOS NAND GATE 86
CMOS NOR GATE 86
CMOS PLL INTEGRATED CIRCUIT 148, 157
CMOS TECHNOLOGY 83
CMRR 111
COIL VOLTAGE 223
COLD SOLDER JOINTS 52, 55, 61, 101, 125, 140, 155, 159, 160, 163, 184, 190, 198, 203, 207, 223
COLLECTOR 36, 38, 51, 169, 197
COLLECTOR CURRENT 50, 75
COLLECTOR VOLTAGE WAVEFORM 31
COLLECTOR-BASE JUNCTIONS 112
COLOR TELEVISION 143
COLPITTS OSCILLATOR 92, 93
COMMAND 217
COMMERCIAL 27
COMMERCIAL APPLICATIONS 30, 144
COMMON GROUND 33
COMMON VOLTAGE 53
COMMON-EMITTER CIRCUIT 74
COMMON-EMITTER STAGE 220
COMMON-EMITTER TRANSISTOR AMPLIFIERS 219
COMMON-MODE GAIN 237
COMMON-MODE INPUT IMPEDANCE 128
COMMON-MODE INPUT RANGE 98
COMMON-MODE INPUT VOLTAGE 237
COMMON-MODE REJECTION 109
COMMON-MODE REJECTION RADIO (CMRR) 111
COMMON-MODE REJECTION RATIO 128, 134, 237
COMMON-MODE SIGNAL 127, 128, 131, 134
COMMUNICATION 144
COMMUNICATIONS PAGING DECODERS 213
COMPARATOR 115, 167-169, 170, 171, 173
COMPARATOR CIRCUIT 115
COMPENSATION 237
COMPLEMENTARY METAL-OXIDE SILICON (CMOS) 83
COMPLEMENTARY OUTPUT SIGNALS 73
COMPLEMENTARY SIGNALS 75
COMPONENT RATING VALUES 20
COMPONENT SELECTION 17

COMPONENT TOLERANCES 131
COMPONENT VALUES 158, 163
COMPONENTS 12, 22, 67, 91, 92, 95, 98,
    101, 115, 124, 131, 132, 133, 134, 148, 158,
    159, 167, 188, 189, 198, 205, 206, 208, 219
COMPUTER 17, 27, 157
COMPUTER GAMES 157
COMPUTER-AIDED POWER SUPPLY 17
CONDITIONERS 133
CONDUCTING CHANNEL 83
CONDUCTING ELECTRODE 83, 84
CONDUCTIVE FOAM 89
CONDUCTIVITY 83
CONNECTIONS 63, 65, 147, 160
CONSTRUCTION 54
CONSTRUCTION ERRORS 52
CONTACT RESISTANCES 134
CONTROL ELEMENT 33
CONTROL FLIP-FLOP 167, 168, 169, 170, 173
CONTROL GATES 148
CONTROL INPUT 84
CONTROL SIGNALS 94
CONTROL VOLTAGE 150
CONTROL VOLTAGE CORRECTION 145
CONTROL VOLTAGE INPUT TERMINAL 147, 150
CONTROL-SWITCH CIRCUIT 198
CONVERSION EFFICIENCY 27
COPPER WIRE 181
COPPER WIRE WINDINGS 181
COULOMBS 12
COUNT SEQUENCE 195
COUNTER 195
COUNTER OUTPUTS 93
COUNTING STAGES 93
COUPLING CAPACITOR 107, 110, 115, 122,
    158, 212, 214
COURSE FOLLOWER 133
CRYSTAL OSCILLATOR REFERENCE 145
CURRENT 8, 11-14, 31, 34, 49, 57, 74-
    77, 83, 84, 88, 89, 94, 108, 110, 157,
    158, 169, 183, 212, 218, 221
CURRENT CAPABILITY 123
CURRENT CLAMPING 237
CURRENT DRAIN 110, 203
CURRENT FLOW 8, 11, 31, 47, 51, 59,
    77, 187, 198, 202, 205, 218, 221, 223
CURRENT GAIN 16, 59, 218
CURRENT HANDLING 27
CURRENT LIMIT (CL) 36
CURRENT LIMIT RESISTORS 59
CURRENT LIMITER TRANSISTOR 36
CURRENT LIMITING 36
CURRENT LIMITING CIRCUIT 34
CURRENT LIMITING FACILITY 36
CURRENT LIMITING POINT 42
CURRENT LIMITING RESISTOR 15, 42, 219
CURRENT LIMITING TRANSISTOR 59
CURRENT OUTPUT 218

CURRENT RATING 22, 49, 171
CURRENT REGULATOR 49
CURRENT RESISTANCE 75
CURRENT SENSE (CS) 36
CURRENT SINKING 168
CURRENT SOURCE 36, 51, 59, 175
CURRENT SOURCING 168
CURRENT TRANSIENTS 30
CURRENT-LIMITED 123
CURRENT-LIMITING RESISTOR 162, 202
CURRENTS 33, 88, 112
CUT-TONE CONTROL CIRCUITS 137
CUTOFF 74
CUTOFF FREQUENCY 115, 132, 136, 139, 214
CUTOFF MODE 76, 77
CYCLE 237
CYCLE TIME 158

D

D FLIP-FLOP 98
DARKROOM TIMER 205, 206, 207, 208
DARLINGTON 59
DARLINGTON PAIR AMPLIFIER 218
DARLINGTON PAIR TRANSISTOR AMPLIFIER
    217, 218
DAT RECORDER 137
DATA 98
DC (DIRECT CURRENT) 5, 12, 108, 150, 162
DC COMPONENT 8, 10, 212, 213
DC DIFFERENTIAL VOLTAGE 110
DC ERROR VOLTAGE 143
DC LOAD VOLTAGE 12
DC MOTOR 181, 183
DC MOTOR CONTROLLER 183
DC MOTOR DRIVE 181
DC MOTOR FIELD WINDING 181
DC MOTOR SPEED 181
DC MOTOR SPEED CONTROL 181
DC MOTOR SPEED CONTROLLER 181, 183, 184
DC OFFSET LEVEL 132
DC OPERATIONAL AMPLIFIER 27
DC OUTPUT 12, 222
DC POTENTIAL 147
DC REFERENCE VOLTAGE 147
DC SIGNALS 122
DC SOURCE 181
DC SUPPLIES 14
DC VOLTAGE 5, 7, 8, 12, 49, 54, 59, 67, 68,
    86, 91, 110, 124, 137, 183, 197, 205, 222
DC VOLTAGE LEVEL 147
DC VOLTAGE POWER SUPPLY 27
DC VOLTAGE SOURCE 5
DC VOLTMETER 156, 199, 204
DC-COUPLED FLIP-FLOP CIRCUIT 115
DECADE COUNTER 93, 94, 195, 196
DECADE SHIFT COUNTER 195
DECIBELS 139, 188

Index

DECIMAL NUMBER 195
DECIMAL TIME SLOT 195
DECODE 217
DECODED OUTPUTS 195
DECODER 94, 119, 120, 122, 123, 125, 219-221, 224
DECODING CIRCUITS 219
DECOUPLING CAPACITOR 212
DELAY 176
DELTA BRAIN WAVES 127
DEMODULATED OUTPUT TERMINAL 147
DESIGN APPLICATIONS 34
DESIGN FORMULAS 10, 12
DESIGN MODE 214
DESIGNER 47
DESIRED FREQUENCY 148
DEVICE 217
DIAGRAM 124
DIFFERENCE INPUT SIGNAL 131
DIFFERENCE SIGNAL 131
DIFFERENTIAL AMPLIFIER 53, 107, 108, 111, 114, 123, 128, 129, 131, 134, 168, 237
DIFFERENTIAL AMPLIFIER INPUT STAGE 107, 110
DIFFERENTIAL AMPLIFIER OUTPUT 128
DIFFERENTIAL GAIN 107, 108
DIFFERENTIAL INPUT CIRCUIT 128
DIFFERENTIAL INPUT IMPEDANCES 128
DIFFERENTIAL INPUT VOLTAGE 111
DIFFERENTIAL INPUTS 108, 128
DIFFERENTIAL VOLTAGE 27
DIFFERENTIAL-MODE GAIN 237
DIFFERENTIAL-MODE INPUT 238
DIFFERENTIATION 114
DIGITAL APPLICATIONS 115
DIGITAL CIRCUITS 97, 153, 156
DIGITAL CONTROL 145
DIGITAL FREQUENCY DIVIDER 146, 147
DIGITAL LOGIC PROBE 97, 98, 99, 101, 102, 103
DIGITAL LOGIC PROBE TIP 102
DIGITAL SIGNAL 167
DIGITAL SYSTEMS 73
DIODE CURRENTS 10
DIODE PRV 22
DIODE RECTIFIER BRIDGE 94
DIODE WAVEFORMS 10
DIODES 7, 8, 10-12, 31, 47, 49, 50, 52, 54, 55, 57, 59, 61, 65, 67, 76, 77, 79, 91, 94, 98, 99, 101, 102, 124, 125, 132, 133, 137, 139, 154, 155, 159, 160, 181, 183, 187, 189, 190, 197, 198, 205, 206, 207, 222, 223
DIRECT COUPLED 108
DIRECT CURRENT (DC) 5
DIRECT CURRENT FLOW 13
DIRECT MODE 132
DIRECT POSITION 131, 135
DIRECT VOLTAGE 61
DIRECT-COUPLED CIRCUITS 110

DISCHARGE 88, 188, 189
DISCHARGE CURRENT 171
DISCHARGE CYCLE 96
DISCHARGE PERIOD 93
DISCHARGE PULSE 93
DISCHARGE TERMINAL 170
DISCHARGE TIME 12, 91, 92, 170
DISCHARGED BATTERY 49
DISCRETE COMPONENTS 27, 31
DISPLAY 95, 96
DISPLAY CIRCUITS 95
DISPLAY DECIMAL POINT 94
DISTORTION 123
DIVIDE-BY-N CIRCUIT 144
DIVIDERS 93
DOLBY LABORATORIES 119
DOLBY SYSTEM 119
DOORBELL TRANSFORMER 189
DRAIN 83, 84
DRAIN-TO-SUBSTRATE JUNCTION 84
DRIFT 134
DRIFT FREE 134
DRILLING TEMPLATE 190
DRIVE 59
DRIVER AMPLIFIER 218
DRIVER CIRCUIT 75
DRIVER STAGE 73
DRIVER TRANSISTOR 75
DRIVERS 94, 95
DRIVING GATE 77
DUAL J-K MASTER-SLAVE FLIP-FLOPS 221
DUAL POTENTIOMETER 122
DUAL POWER SUPPLY 20, 53, 109, 110, 146, 147, 155
DUAL SUPPLY 154
DUAL-D FLIP-FLOPS 93, 98
DUAL-IN-LINE PACKAGE 211
DUAL-PHASE OUTPUTS 93
DUAL-TRACKING POWER SUPPLY 67, 68
DUTY CYCLE 102, 148, 149, 163, 167, 170, 181, 183, 238
DUTY CYCLE CONTROL 31
DUTY CYCLE OSCILLATOR 176

E

EARLOBE 135
EDGE-TRIGGERED FLIP-FLOPS 148
ELECTRIC VEHICLES 181
ELECTRICAL ACTIVITY 127
ELECTRICAL CONTACT 135
ELECTRICAL SIGNALS 107
ELECTROCARDIOGRAM (EKG) 127
ELECTRODE CREAM 134, 135
ELECTRODE PLACEMENTS 135
ELECTRODE POTENTIALS 128
ELECTRODES 84, 133-135
ELECTROENCEPHALOGRAM (EEG) 127

IC Design Projects

ELECTROLYTIC CAPACITOR 65, 133
ELECTROMYOGRAM (EMG) 127
ELECTRONIC CIRCUITS 5, 12, 27, 28, 67, 153
ELECTRONIC COMPONENTS 111
ELECTRONIC ENGINEERS 107
ELECTRONIC ORGANS 185, 187, 189, 192
ELECTRONIC SWITCH 181
ELECTRONS 83, 132
EMITTER 36, 38, 74, 76
EMITTER CURRENT 50, 51, 76
EMITTER TERMINALS 51
EMITTER VOLTAGE 16
EMITTER-FOLLOWER CIRCUIT 74
ENCLOSURE 201
ENCODE 217
END CAP 101, 102
END PLATE 102
ENERGY 14, 133
ENLARGER 205, 206, 208
ERROR AMPLIFIER 33, 36
ERROR AMPLIFIER INPUTS 33
ERROR AMPLIFIER OUTPUT 39
ERROR SIGNAL 59, 145
ERROR VOLTAGE 143, 145
EUROPE 5
EUROPEAN EMERGENCY VEHICLES 201
EXCLUSIVE-OR GATE 149
EXCURSION 113
EXPERIMENTER 129
EXPLOSIONS 120, 121
EXPOSURE TIME 208
EXTERNAL CAPACITOR 39, 146, 147, 212
EXTERNAL CIRCUIT 150
EXTERNAL COMPONENTS 33, 113, 211
EXTERNAL FIELDS 129
EXTERNAL INTERFERENCE 129
EXTERNAL INTERFERENCE VOLTAGES 128
EXTERNAL NOISE 129
EXTERNAL RESISTOR 146, 147, 170, 212
EXTERNAL TIMING CAPACITOR 169, 212
EXTERNAL TIMING COMPONENTS 171, 195
EXTERNAL TIMING RESISTOR 169, 212
EXTERNAL TRANSISTORS 27
EYEBROW 135
EYES 135

F

FALSE SIGNALS 212
FALSE TRIGGERING 173
FAN-IN 79, 87
FAN-OUT 77, 79
FARADS 12
FEEDBACK 31, 110, 133
FEEDBACK COMPONENT 115
FEEDBACK LOOP 108, 109, 144
FEEDBACK MONITOR 133
FEEDBACK NETWORK 107, 109
FEEDBACK RESISTORS 138
FIELD CURRENT 181
FIELD-EFFECT TRANSISTORS (FET) 109
FIELDS 128
FILTER 115, 136, 144, 212, 214
FILTER CAPACITORS 59, 154, 155
FILTERING 14, 17, 30
FIXED OUTPUT VOLTAGE 28
FIXED RESISTANCE 91
FIXED-VALUE RESISTORS 192
FIXED-VOLTAGE OUTPUT 54
FIXED-VOLTAGE REGULATORS 32, 53, 54
FIXED-VOLTAGE SUPPLIES 42
FLAT OPERATION 111
FLATS 185
FLIP-FLOP 93, 98, 99, 115, 221, 237, 238
FLIP-FLOP OUTPUT 170, 173
FLOATING REGULATOR 47
FLUCTUATING DC VOLTAGE 54, 67, 205
FLUCTUATING DC VOLTAGES 154
FLUCTUATIONS 57
FLUX 6
FM DETECTION 148
FM DETECTOR CIRCUIT 147
FM DISCRIMINATORS 147
FOCUS POSITION 205, 208
FOLDBACK CURRENT 63
FOLDBACK CURRENT LIMITING 57, 59
FOLDBACK CURRENT-LIMITING CIRCUIT 42, 47
FOLDBACK LATCHUP 62
FOLDBACK LIMITING 42
FORMULAS 8, 10, 12, 15, 16
FORWARD VOLTAGE DROP 80
FORWARD-BIASED 74
FORWARD-BIASING DIODES 99
FREE-RUNNING FREQUENCY 214
FREE-RUNNING OSCILLATOR 115, 169, 202, 237
FREQUENCY 8, 10, 12, 91, 103, 108, 109, 111, 115, 120-122, 127, 129, 132, 137-139, 143-146, 148-150, 156, 158, 162, 164, 167, 176, 185, 188, 189, 202, 220, 223, 238
FREQUENCY ADJUST CONTROL 153
FREQUENCY COMPENSATION 39
FREQUENCY CONTROL 153, 213
FREQUENCY CORRECTIONS 145
FREQUENCY COUNTER 193
FREQUENCY DATA 193
FREQUENCY DETECTION BANDWIDTH 214
FREQUENCY DIVIDER 173
FREQUENCY LIMIT 132
FREQUENCY MONITORING 213
FREQUENCY MULTIPLICATION 146, 147
FREQUENCY OF OPERATION 153, 162, 170, 218, 220
FREQUENCY OF OSCILLATION 143, 149, 158, 176
FREQUENCY RANGE 129, 135, 153, 212
FREQUENCY RESPONSE 126, 139, 188
FREQUENCY SYNTHESIZERS 144, 145

Index

FREQUENCY-SHIFT KEYING (FSK) 147
FRONT CHANNEL 121
FRONT CHANNEL SIGNALS 122
FRONT SPEAKER 119, 120
FULL-WAVE BRIDGE CENTER-TAPPED CIRCUIT 17
FULL-WAVE BRIDGE CIRCUIT 17
FULL-WAVE BRIDGE LOAD VOLTAGE 12
FULL-WAVE BRIDGE RECTIFIER 7, 10, 12, 13
FULL-WAVE BRIDGE RECTIFIER CIRCUIT 10
FULL-WAVE CENTER-TAPPED CIRCUIT 17
FULL-WAVE RECTIFIER 7, 8, 13, 49
FULL-WAVE RECTIFIER CIRCUIT 8, 10
FULL-WAVE RECTIFIER VOLTAGE 10
FULL-WAVE RECTIFIER WAVEFORMS 10
FULL-WAVE VOLTAGE 10
FUNCTION 113
FUNCTION GENERATOR 153, 154, 155, 156
FUNCTION GENERATOR CIRCUIT 153
FUSE 199, 206
FUSE HOLDER 199, 207

G

GAIN 85, 107, 108, 111, 121, 122, 126, 132, 134, 136-139, 147, 220
GAIN CONTROL 120, 123, 126, 132, 153
GAIN RESPONSES 109
GAIN-BANDWIDTH PRODUCT 111, 220
GATE 74, 75, 78, 83, 84, 87
GATE DELAY 80, 88
GATE INPUT 88
GATE INPUT CURRENT 76
GATE RELAY 79
GATE VOLTAGE 83
GATE-TO-SUBSTRATE POTENTIAL 84
GREEN 102, 163, 164
GROOVE MODULATION 137
GROUND 36, 44, 47, 54, 63, 75, 85, 88, 89, 102, 113-115, 146, 150, 155, 159, 163, 164, 169, 170, 199, 204, 212-214
GROUND ELECTRODE 135
GROUND LEAD 102
GROUND POTENTIAL 86, 134, 135, 150
GROUND VOLTAGE 53
GROUNDED ASSEMBLY TABLES 89
GROUNDED TIP 159, 163

H

HALF-WAVE RECTIFIED VOLTAGE 10
HALF-WAVE RECTIFIER 7, 8, 13
HALF-WAVE RECTIFIER CIRCUIT 8, 10
HALF-WAVE RECTIFIER WAVEFORMS 8
HAND 135
HARMONIC DISTORTION 188
HARMONICS 148, 149
HEAD 128, 135
HEART 127
HEARTBEAT 135
HEAT 52, 60
HEAT DAMAGE 101, 163, 203
HEAT SINK 52, 55, 60, 183, 184, 190
HEAT SINKING 30, 33
HEAT SINKS 68
HEAVY DISCHARGE 218
HEIL 83
HEX BUFFER 196
HIGH LEVELS 97
HIGH-CURRENT TRANSISTORS 27
HIGH-FREQUENCY APPLICATIONS 111
HIGH-FREQUENCY LIMIT 132
HIGH-FREQUENCY SIGNALS 123
HIGH-FREQUENCY SOUNDS 123
HIGH-GAIN AMPLIFIER 219
HIGH-GAIN INVERTING AMPLIFIER 98
HIGH-PASS 115
HIGH-PASS FILTER 115, 132, 139
HIGH-SPEED TTL 80
HIGH-SPEED TTL GATES 79
HOBBYIST 101
HOBBYISTS 27
HOLE 102
HOLES 190, 199, 207
HOLIDAY 195
HOME 157, 199, 224
HOME THEATER SYSTEM 119, 120, 121
HOUSE 195
HOUSEHOLD AC LINE VOLTAGE 67, 137, 154, 159, 197, 205, 222
HOUSEHOLD LINE VOLTAGE 5, 8, 10, 12, 57, 91, 124
HOUSEHOLD VOLTAGE 8
HOUSEHOLDS 119
HUM 128, 129
HUMAN BEINGS 127, 129
HUMAN BODY 128, 129
HUMAN HEAD 128

I

I-PHASE DETECTOR 211, 212
IC 31, 57, 61, 73, 88, 98, 123-125, 133, 139, 154, 155, 159, 160, 163, 167, 183, 190, 195, 196, 198, 201, 203-205, 207, 208, 221, 223
IC OP-AMP 107, 111, 112
IC REGULATOR 27, 32, 68
IC SOCKETS 61, 101, 102, 133, 139, 154, 159, 163, 198, 203, 207, 223
IC STEREO PREAMPLIFIER 137, 140
IC TRANSISTORS 112
IC VOLTAGE REGULATOR 59, 137
IMPEDANCE 108, 109, 111, 121, 128, 129, 153, 162, 206
IMPEDANCE DIFFERENTIAL AMPLIFIER 129
IMPEDANCE LOADS 123
IMPEDANCE PATH 110
IMPEDANCE SOURCE 150

IN-LINE FUSE 199
INCANDESCENT BULB 184
INDICATOR 187
INDUCTIVE KICKBACK 238
INDUCTIVE SPIKE VOLTAGES 197, 206, 222
INDUCTIVE-SWITCHING DAMAGE 206
INDUCTOR 14
INDUCTOR CURRENT 31
INDUSTRIAL APPLICATIONS 30
INFORMATION 121
INFRARED BEAM 225
INFRARED BURGLAR ALARM SYSTEM 224
INFRARED LEDS 217, 218, 219
INFRARED PHOTOTRANSISTOR 219
INFRARED SIGNALS 219
INFRARED-EMITTING DIODE 184
INHIBIT INPUT TERMINAL 149
INPUT 29, 36, 74-77, 79, 86, 87, 88, 98, 99,
    107, 109, 111, 113-115, 126, 128, 129, 133-
    135, 139, 147, 150, 162, 168, 188, 212
INPUT AMPLIFIER 120, 122
INPUT BASE BIAS CURRENTS 110
INPUT BIAS CURRENT 110, 111, 238
INPUT CAPACITANCE 88
INPUT CAPACITOR 14
INPUT CIRCUIT 73, 110
INPUT CIRCUIT OPERATION 74
INPUT CIRCUITRY 95
INPUT CURRENT 75, 108
INPUT DC VOLTAGE 68
INPUT DEVICE 88
INPUT DIFFERENTIAL AMPLIFIER 107
INPUT DIFFERENTIAL AMPLIFIER STAGE 109, 110
INPUT DIODES 88
INPUT DRIVE CURRENT 74
INPUT FREQUENCY 143, 144, 211
INPUT IMPEDANCE 109, 111, 113, 121, 128, 129,
    131, 150, 153, 218, 220
INPUT OFFSET CURRENT 110, 111, 238
INPUT OFFSET VOLTAGE 110, 111, 138, 238
INPUT PULSE 148, 169, 171, 176
INPUT PULSE FREQUENCY 99
INPUT PULSE TRAIN 176
INPUT RESISTANCE 88
INPUT SIGNAL 79, 111-114, 120, 123, 143, 144,
    146-149, 168, 169, 211, 212, 214
INPUT SIGNAL AMPLIFIER 150
INPUT SIGNAL FREQUENCY 143, 144, 212
INPUT SIGNAL VOLTAGE 147
INPUT STAGE TRANSISTORS 112
INPUT TERMINAL 27, 47, 107, 110, 146, 149, 169,
    213
INPUT TRIGGER PULSE 171
INPUT TRIGGER SIGNAL 115
INPUT VOLTAGE 17, 31, 33, 34, 54, 59,
    68, 83, 88, 98, 99, 108, 110, 113-115
INPUT VOLTAGE RANGE 47
INPUT VOLTAGES 33, 107, 108, 112

INSTRUMENTATION AMPLIFIER 128, 129, 131, 134
INSTRUMENTATION AMPLIFIER INPUTS 134
INSULATING BUSHES 52
INTEGRATE MODE 132
INTEGRATE POSITION 131, 135
INTEGRATED CIRCUIT (IC) 1, 27, 60, 92-95, 98,
    101, 102, 107, 120, 133, 137, 138,
    147, 156, 157, 161, 162
INTEGRATED CIRCUIT OP-AMPS 115
INTEGRATED CIRCUIT REGULATORS 27, 43
INTEGRATED VOLTAGE REGULATORS 32
INTEGRATION 114
INTEGRATOR 114, 132
INTERFACE 196
INTERMEDIATE FREQUENCY 148
INTERNAL CURRENT LIMITING 32
INTERNAL DISCHARGE TRANSISTOR 171, 173
INTERNAL DISSIPATION 28
INTERNAL HEATING 28
INTERNAL PROTECTION CIRCUIT 34
INTERNAL RESISTORS 168
INTERNAL THERMAL OVERLOAD 188
INTERNALLY COMPENSATED 121
INTERVALS 221
INVERSE ACTIVE MODE 75
INVERSE OPERATIONS 114
INVERTED INTEGRAL 114
INVERTED OUTPUTS 94
INVERTER 77, 78
INVERTER GATE 78
INVERTER INPUT 85
INVERTER SWITCH 183
INVERTER-SWITCHES 221
INVERTING AMPLIFIER 98, 113, 139, 238
INVERTING INPUT 31, 39, 109, 113, 115, 123,
    134, 138
INVERTING INPUT TERMINALS 113, 114
INVERTING INPUTS 125
INVERTING STAGES 139, 224
INVERTING TERMINAL 113
ISOLATION JUNCTIONS 112
ISOLATION TRANSFORMER 7

J

JOHNSON BINARY CODE 195
JOHNSON DECADE COUNTER 195
JUMPER 61
JUNCTION 171, 176

K

KEY 115
KEYBOARD 190, 192

L

L + R 120
L-R DECODER 120, 123

Index

L-R REAR CHANNEL SIGNAL 122
L-REGULATOR 14
LAMP CONTROL-SWITCH CIRCUITS 198
LAMP DIMMER 184
LAMPS 197-199
LATCH 224
LATCHING FUNCTIONS 221
LATCHUPS 121
LC FILTER 31
LEADING 98
LEADS 57, 89, 135
LEAKAGE 96
LEAKAGE CURRENTS 88
LED 96, 97, 102, 103, 161, 205, 207, 208, 219, 223
LEFT CHANNEL 124
LEFT CHANNEL SIGNAL 120
LEFT INPUT SIGNAL 120
LEVEL SHIFTER 107
LEVEL-SHIFTER STAGE 107, 108
LIGHT 195
LIGHT-EMITTING DIODE (LED) 95, 97-99, 102, 187, 205
LIGHTS 195
LILIENFELD 83
LIMITED-ADJUSTABILITY SUPPLIES 42
LINE REGULATION 34, 36
LINE TRANSIENTS 30
LINE VOLTAGE 8, 12, 91, 124
LINEAR AMPLIFIER 79
LINEAR CIRCUITS 107, 113
LINEAR FREQUENCY 145
LINEAR IC BUILDING BLOCK 107
LINEAR IC REGULATOR 27
LINEAR ICS 53, 167
LINEAR OP-AMP CIRCUITS 115
LINEAR OPERATION 111
LINEAR POWER SUPPLY 27
LINEAR RAMP 175
LINEAR RAMP GENERATOR 175
LINEAR RANGE 113
LINEAR REGULATOR 27, 28, 30
LINEAR SIGNAL 167
LISTENER 123
LISTENING ROOM 119, 120, 126
LISTENING ROOM ACOUSTICS 137
LIVE PERFORMANCE 119
LIVE PERFORMANCES 123
LIVING CELLS 127
LIVING ORGANISMS 127, 129
LM317 43, 44, 47, 67
LM317 ADJUSTABLE REGULATOR 43
LM317 REGULATORS 68
LM324 98
LM337 43, 47, 67
LM337 ADJUSTABLE REGULATOR 43
LM337 REGULATORS 68
LM340 34

LM383 123
LM383 INTEGRATED CIRCUITS 124
LM555 162, 167, 189, 205
LM555 TIMER 171, 173, 176, 177, 189, 195, 205, 218
LM555 TIMER ASTABLE MULTIVIBRATOR 176, 181, 183
LM555 TIMER IC 92, 162, 167, 168, 169, 173, 183, 195, 201, 205, 206, 217, 218, 220
LM555 TIMER PACKAGE 167
LM555 TIMER PULSE POSITION MODULATOR 175
LM555 TIMER PULSE-WIDTH MODULATOR 173
LM565 147, 148
LM565 PLL 147
LM565 PLL INTEGRATED CIRCUIT 145, 147
LM567 211, 212, 220
LM567 CIRCUIT 211
LM567 TONE DECODER 211, 212, 213, 214
LM567 TONE DECODER IC 211, 212
LM723 36, 38, 39, 42, 57, 59
LM723 IC VOLTAGE REGULATOR 36
LM723 REGULATOR 36
LM741 121
LM741 OP-AMPS 124, 136
LOAD 6, 7, 10, 11-14, 36, 45, 47, 57, 60, 62, 63, 65, 67, 76, 77, 184, 221, 224
LOAD CAPACITANCES 88
LOAD CHARACTERISTICS 181
LOAD CIRCUIT 63, 65
LOAD CONDITIONS 59
LOAD CURRENT 8, 14, 16, 17, 28, 30, 31, 36, 46, 47, 49, 59, 60, 77
LOAD IMPEDANCE 49, 77, 108, 188
LOAD REGULATION 27, 34, 36
LOAD SUMS 10
LOAD TRANSIENTS 30
LOAD VOLTAGE 8, 10, 12, 14, 15, 60
LOCK RANGE 143, 147
LOCOMOTIVES 181
LOG AMPLIFIER 116
LOGIC CIRCUITS 221
LOGIC COMPATIBLE OUTPUT 212
LOGIC FAMILIES 73, 89
LOGIC FUNCTIONS 79
LOGIC LEVEL 97, 99, 161
LOGIC LEVEL MEASUREMENT 97
LOGIC PROBE 97-99, 101, 102, 161-164
LOGIC PROBE TIP 102, 103, 161, 164
LONG MODE 206
LONG POSITION 205
LONG-TERM VOLTAGES 134
LOOP 147, 148
LOOP FILTER 144, 145, 213
LOOP-FILTER CAPACITOR 148, 214
LOUDSPEAKER 133, 158, 162, 163, 204
LOUDSPEAKER IMPEDANCE 158
LOW 212
LOW IMPEDANCE 133

IC Design Projects

LOW IMPEDANCE DC PATH 110
LOW LEVELS 97
LOW STATE 173
LOW-FREQUENCY INFORMATION 120
LOW-FREQUENCY LIMIT 132
LOW-FREQUENCY RESPONSE 120
LOW-LEVEL SIGNAL 98
LOW-PASS 115
LOW-PASS FILTER 12, 115, 120, 122, 123, 132, 133, 139
LOW-PASS LOOP FILTER 143, 144, 147, 212, 213, 214
LOW-PASS LOOP FILTER CAPACITOR 214
LOW-PASS LOOP FILTER RESISTOR 147
LOW-PASS OUTPUT FILTER 212, 213
LOWER COMPARATOR 176
LOWER VOLTAGE 31
LSI (LARGE-SCALE INTEGRATION) 73, 89
LSI PACKAGES 79
LVSI PACKAGES 89

M

MAGNETIC CARTRIDGE 138
MAGNETIC CORE 6
MAGNETIC PHONO CARTRIDGE 137
MAJORITY CARRIERS 83
MANUFACTURERS 85
MANUFACTURING COSTS 28
MASTER-SLAVE FLIP-FLOPS 221
MATERIALS 83
MATHEMATICAL OPERATIONS 107
MATHEMATICS 114
MAXIMUM DUTY CYCLE 171
MAXIMUM VOLTAGE 5
MC7800 32, 33, 34, 43, 53
MC7800 REGULATORS 32
MC78XX 54, 55
MC7900 32, 33, 34, 43, 53
MC7900 REGULATORS 32
MC79XX 54, 55
MEASUREMENT 133, 134
MEASUREMENT EQUIPMENT 144
MEASURING DEVICE 97
MEASURING INTERVAL 93
MEMORY 89, 97, 98, 99
MEMORY OUTPUT DEVICE 98
MENTAL ACTIVITY 127
MENU 22
METAL ENCLOSURE 184, 199, 207
METAL GATE 88
METAL OXIDE SILICON (MOS) 83
METER 57, 91, 94, 95
MICA INSULATOR 60
MICA WASHER 52, 60, 184
MICROAMPERES 46
MICROFARADS 173, 214
MICROPROCESSOR ICS 89

MINI-DIP PACKAGE 167
MISSILE TRACKING 143
MISSING PULSE 177
MISSING PULSE DETECTOR 176
MISSING PULSE DETECTOR CIRCUITS 167
MIXER CIRCUITS 189
MODE OF OPERATION 167
MODE SWITCH 135
MODULATING SIGNAL 173
MONITOR 131
MONITOR CIRCUIT 136
MONOLITHIC CHARACTERISTICS 27
MONOLITHIC CIRCUITS 27
MONOLITHIC COMPONENTS 27
MONOLITHIC CONSTRUCTION 28
MONOPHONIC 120
MONOPHONIC ELECTRONIC ORGAN 188, 190
MONOPHONIC ORGAN 185, 189, 190
MONOPHONIC SIGNAL SOURCE 139
MONOSTABLE 171
MONOSTABLE CIRCUIT 171
MONOSTABLE MODE 167
MONOSTABLE MULTIVIBRATOR 93, 115, 171, 173, 175, 205, 220, 224, 238
MONOSTABLE MULTIVIBRATOR CIRCUIT 173
MONOSTABLE OPERATION 171
MOS 83
MOS DEVICES 83, 88
MOS FIELD EFFECT TRANSISTOR 83
MOS TRANSISTOR 83
MOSFET 84
MOTION ARTIFACTS 134
MOUNTING HOLES 207
MOUNTING SOCKET 60
MOVIE THEATER 119
MSI (MEDIUM SCALE INTEGRATION) 79, 89
MULTI-EMITTER TRANSISTOR 78
MULTIDECADE COUNTER CHAIN 196
MULTILEAD POWER PACKAGES 27
MULTIVIBRATOR 133
MULTIVIBRATOR CIRCUIT 115
MUSCLES 127, 129
MUSIC 192
MUSIC BALANCE 137
MUSICAL NOTE 185, 188, 189, 190, 191
MUSICAL NOTE GENERATOR CIRCUITS 189
MUSICAL RANGE 185

N

N-CHANNEL 84, 85
N-CHANNEL DEVICE 83-86
N-REGIONS 83
N-TRANSISTOR 83
N-TYPE SILICON 83, 84
NAND GATE 77, 86, 157
NAND GATE ASTABLE MULTIVIBRATOR 157
NANOWATTS 87

Index

NEEDLE 102, 204
NEGATIVE FEEDBACK 31, 113
NEGATIVE FEEDBACK LOOP 138
NEGATIVE FEEDBACK OP-AMP CIRCUIT 109
NEGATIVE PULSE 173
NEGATIVE RAIL 146, 147
NEGATIVE SATURATION 115
NEGATIVE SUPPLY VOLTAGE 115
NEGATIVE TEST LEAD 199, 204
NEGATIVE VOLTAGE REGULATOR 32, 38, 43, 68
NERVES 127, 129
NERVOUS SYSTEM SENSORS 127, 129
NICKEL-CADMIUM RECHARGEABLE BATTERIES 49
NMOS TRANSISTOR 83, 87
NO-LOAD OUTPUT VOLTAGE 12
NOISE 79, 111, 128, 129, 131, 134, 136, 149, 212
NOISE DISPLAY 135
NOISE IMMUNITY 83, 85
NOISE MARGIN 79, 87
NOISE REJECTION 131
NOISE SPIKES 134, 144
NOISE VOLTAGE 87, 111
NON-INVERTING AMPLIFIER 98
NON-PORTABLE EQUIPMENT 5
NONINVERTING AMPLIFIER 113, 120, 138, 188, 238
NONINVERTING INPUT 31, 36, 123, 134, 138, 153
NONINVERTING INPUT TERMINAL 113, 114, 115
NONINVERTING INPUTS 125
NONINVERTING STAGE 139
NONINVERTING TERMINAL 113
NONLINEAR CIRCUITS 115
NONLINEAR ELECTRONIC CIRCUITS 107
NONLINEAR OP-AMP CIRCUITS 116
NONLINEAR RANGE 113
NOR GATE 86
NORMALLY-OFF AC SOCKET 208
NORMALLY-ON AC SOCKET 208
NORTH AMERICA 5
NOTCH FILTER 115, 136
NOTES 185
NPN SERIES-PASS TRANSISTOR 36
NPN TRANSISTORS 36
NULL STAGE 131, 136

O

OCTAVES 132, 185, 190, 191
OFF STATE 171
OFFICE 224
OFFSET 134
OFFSET CURRENT 138
OFFSET VOLTAGE 110, 111, 134
OHMS 46, 173
ONE LEVEL 85
ONE-SHOT MULTIVIBRATOR 116, 171
ONSCREEN DISPLAY 215

OP-AMP 98, 107, 108-115, 124, 131, 132, 136-139, 156
OP-AMP CIRCUITS 115, 116
OP-AMP DIFFERENTIATOR 114
OP-AMP INPUT 111
OP-AMP INTEGRATOR 114
OP-AMP OUTPUT VOLTAGE 111
OP-AMP STAGES 111
OPEN CIRCUIT 122, 162, 164
OPEN-COLLECTOR TRANSISTOR 169
OPEN-LOOP CIRCUIT 115
OPEN-LOOP GAIN 108, 109, 113, 238
OPERATING CURRENT 46, 47
OPERATING PROBLEMS 101
OPERATION 162
OPERATIONAL AMPLIFIER 27, 31, 98, 107-109, 121, 123, 131, 133, 153, 220, 238
OPERATIONAL AMPLIFIER CIRCUITS 53, 113
OPERATIONAL PROBLEMS 125
OPTOISOLATOR DEVICE 135, 136
OPTOISOLATORS 129
ORGAN 185, 189, 192
OSCILLATE 150
OSCILLATION 109, 143, 146, 148, 169, 213
OSCILLATION FREQUENCY 143, 147
OSCILLATOR 95, 167, 187, 201, 213, 238
OSCILLOSCOPE 129, 134-136, 155, 193
OUT OF PHASE 113
OUT-OF-PHASE OUTPUT SIGNALS 73
OUTPUT 27, 29, 31, 40, 44, 47, 59, 75, 77, 79, 87, 93, 102, 109-115, 125, 135, 137, 139, 145, 147, 148, 150, 157, 162, 167-169, 173, 176, 183, 187-189, 191, 202, 211, 212, 214, 220
OUTPUT BUFFER STAGE 167, 168
OUTPUT CIRCUIT 73, 74
OUTPUT CIRCUIT OPERATION 74
OUTPUT CURRENT 31, 36, 57, 59, 62, 63
OUTPUT CURRENT CAPABILITY 27
OUTPUT CURRENT FLOW 36
OUTPUT DECODER 195
OUTPUT DELAY 211
OUTPUT DEVICE 20
OUTPUT FILTER 212
OUTPUT FREQUENCY 102, 145, 202
OUTPUT IMPEDANCE 108, 109, 111, 113, 121, 153, 218
OUTPUT LEVEL 177
OUTPUT OFFSET 138
OUTPUT OFFSET VOLTAGE 110, 239
OUTPUT POWER 219
OUTPUT POWER AMPLIFIER STAGE 107
OUTPUT PULSE 169, 170, 197
OUTPUT PULSE WIDTH 173
OUTPUT RESISTANCE 113
OUTPUT RIPPLE 31
OUTPUT RIPPLE FACTOR 30
OUTPUT SIGNAL 113, 114, 133

OUTPUT SIGNAL CONFIGURATION 75
OUTPUT SIGNALS 113, 202
OUTPUT STAGE 112, 148
OUTPUT STATE 168
OUTPUT TERMINAL 46, 47, 57, 89, 107, 108, 146, 213
OUTPUT TRANSISTOR 76
OUTPUT VOLTAGE 27, 30-34, 36, 38, 42, 46, 47, 49, 54, 57, 59, 61-63, 68, 69, 74-77, 85, 86, 107, 108, 110, 111, 113, 115, 145, 147
OUTPUT VOLTAGE OPTIONS 34
OUTPUT VOLTAGE RANGE 47
OUTPUT VOLTAGE REGULATION 16
OUTPUT VOLTAGE SWING 88
OUTPUT WAVEFORM 156, 163, 167, 202
OUTPUTS 93, 98, 99, 189, 195, 220
OVERFLOW CONDITION 95
OVERFLOW PULSES 93, 94
OVERLOADS 31, 57
OVERRANGE INDICATORS 94
OVERRANGE LEDS 96
OVERVOLTAGE PROTECTION 57
OXIDE INSULATOR 88

P

P-CHANNEL 84, 85
P-CHANNEL DEVICE 84-86
P-CHANNEL TRANSISTOR 83
P-DEVICE 84
P-REGIONS 84
P-TRANSISTOR 83, 84
P-TYPE SILICON 83, 84
PACKAGES 47, 123
PAGING DECODERS 213
PARALLEL 12, 14, 17, 86
PARALLEL OPERATION 65
PART NUMBERS 139
PARTS 162, 163
PARTS LIST 51, 54, 60, 68, 94, 101, 125, 133, 139, 154, 159, 163, 164, 183, 190, 198, 203, 207, 223
PASS TRANSISTOR CONTROL ELEMENT 34
PASSBAND 120, 211
PATIENT 129
PEAK-TO-PEAK AMPLITUDE 112
PEAK-TO-PEAK RIPPLE VOLTAGE 12
PELLET 134
PERFBOARD 51, 54, 60, 68, 94, 101, 124, 133, 139, 154, 155, 159, 163, 183, 184, 189, 198, 203, 207, 223
PERFORMANCE 113
PERIOD 239
PERIODIC 239
PHASE COMPARATOR 143, 148, 149, 150
PHASE COMPARATOR VCO INPUT 147
PHASE COMPARISON 145

PHASE DETECTOR 145, 146
PHASE DETECTOR OUTPUT 145
PHASE DISTORTION 139
PHASE INVERSION 109
PHASE-LOCKED LOOP (PLL) 143, 239
PHASE-SPLITTER CIRCUIT 75
PHONO 139
PHONO CARTRIDGE 138
PHONO INPUT 139
PHOTOGRAPHIC PAPER 208
PI-REGULATOR 14
PICOFARADS 95
PIN 36, 38, 95, 98, 125, 146-151, 153, 155, 157, 162, 164, 169, 171, 173, 176, 188, 199, 205, 211-214, 224
PIN ASSIGNMENT 36, 195, 196, 211
PIN CONFIGURATION 124, 136
PIN-FOR-PIN COMPATIBLE 123
PITCH 189
PITCH CONTROL 189
PITCH CONTROL CIRCUITS 185
PITCH CONTROLS 193
PLASTIC 199, 207
PLASTIC TUBE 101
PLL 143, 144, 145
PLL BANDWIDTH 147
PLL CIRCUIT 143, 144, 155
PLL CMOS INTEGRATED CIRCUIT 161, 163
PLL FREQUENCY ACCURACY 145
PLL ICS 144, 145, 147, 148, 153, 157
PLL SYNTHESIZER 145
PLL SYSTEMS 143
PLUG-IN TIMES 200
PMOS TRANSISTOR 83, 87
POINT-TO-POINT WIRING 124
POLARITIES 8, 52, 67
POLARIZATION 134
POLICE SIREN 157
POLICE SIREN EFFECT 158
POLYPHONIC ELECTRONIC ORGAN 189, 190
POLYPHONIC ORGAN 185, 190, 191
PORTABLE UNIT 217
POSITIVE 10
POSITIVE FEEDBACK 113, 133
POSITIVE FEEDBACK LOOP 109
POSITIVE GATE VOLTAGE 83
POSITIVE POWER SUPPLY 151
POSITIVE POWER SUPPLY RAIL 150
POSITIVE RAIL 86, 147, 157, 164, 169, 173, 214
POSITIVE SATURATION 115
POSITIVE SUPPLY VOLTAGE 115
POSITIVE TERMINAL 223
POSITIVE TEST LEAD 199, 204
POSITIVE VOLTAGE REGULATOR 34, 43, 47, 68
POSITIVE VOLTAGE REGULATORS 32
POTENTIOMETER 47, 61-63, 69, 95, 123, 124, 134, 139, 153, 155, 156, 183, 184, 195, 205, 206-208, 224, 239

Index

POWER 16, 29, 30, 61, 87, 98, 181, 198, 204, 205, 224, 225
POWER AMPLIFIER 107, 108, 120, 123, 158, 188
POWER CONSUMPTION 83, 149, 151
POWER CONVERTER 29
POWER DISSIPATION 27, 31, 85
POWER DISTRIBUTION EQUIPMENT 5
POWER DISTRIBUTION FREQUENCY 5
POWER DRIVE SWITCH 219, 221
POWER FAILURE 195, 198
POWER LEADS 60, 99, 102
POWER LINE INTERFERENCE 128
POWER LOAD 7
POWER LOSSES 181
POWER RAIL 63
POWER RATING 7
POWER RESISTOR 59
POWER SUPPLIES 5
POWER SUPPLY 5, 12, 14, 17, 28, 33, 36, 38, 53, 55, 57, 59, 60, 61, 63, 65, 67, 68, 78, 84, 86, 91, 94, 98, 102, 111, 115, 121, 133, 154, 157, 162-164, 169, 173, 187-189, 208, 212, 214, 219, 222, 223
POWER SUPPLY CABINET 55, 60, 68
POWER SUPPLY CIRCUIT 12, 20, 94
POWER SUPPLY CIRCUITS 17
POWER SUPPLY DESIGNER 32
POWER SUPPLY DUAL VOLTAGES 155
POWER SUPPLY LEADS 125
POWER SUPPLY RAIL 85, 88, 102, 155
POWER SUPPLY REGULATION 148
POWER SUPPLY VOLTAGE 12, 77, 85-87, 102, 110, 112, 123, 169, 171, 173, 183, 188, 203
POWER SUPPLY VOLTAGE FLUCTUATIONS 111
POWER SUPPLY VOLTAGE REJECTION RATIO (PSRR) 111
POWER SWITCH 61, 133
POWER TRANSISTOR 60, 183, 184
POWER-ON INDICATOR 219, 223
PREAMPLIFIER 137, 139
PRECISION RECTIFIERS 116
PRECISION TIMING CIRCUITS 167
PRIMARY 239
PRIMARY CURRENT 6
PRIMARY VOLTAGE 6, 7
PRIMARY WINDING 6
PRINTED CIRCUIT BOARD 54, 60
PRINTER 20
PROBE MEMORY 103
PROBE TIP 161, 162
PROBES 133
PROGRAM 17, 22, 173, 214
PROGRAM LISTING 17
PROGRAM OPERATOR 20
PROGRAMMABLE DIVIDER 145
PROGRAMMING 195
PROGRAMMING CURRENT 44, 47
PROJECT 94, 157, 159
PROPAGATION DELAY 79, 87
PRV (PEAK REVERSE VOLTAGE) 8
PRV RATING 10, 12
PSRR 111
PSYCHOPHYSIOLOGICAL PHENOMENA 133
PULL-IN RANGE 146
PULL-UP RESISTOR 175, 189, 212, 214
PULL-UP TRANSISTOR 75
PULSATING DC VOLTAGE 5, 7, 49, 54, 67, 91, 124, 137, 154, 159, 187, 197, 205, 222
PULSATING ERROR VOLTAGE 143
PULSE 99, 103, 162, 164, 171
PULSE DETECTOR 176
PULSE DETECTOR CIRCUITS 167
PULSE FREQUENCIES 99
PULSE POSITION 173
PULSE STREAM 97, 99, 102
PULSE TRAIN 93, 103, 173
PULSE WAVEFORM 31
PULSE WIDTH 93, 99, 115, 205, 206
PULSE-GENERATING CIRCUITS 167
PULSE-POSITION MODULATION 239
PULSE-POSITION MODULATION CIRCUITS 167
PULSE-POSITION MODULATOR 173
PULSE-WIDTH MODULATION 239
PULSE-WIDTH MODULATION CIRCUITS 167
PULSE-WIDTH MODULATOR 173
PULSES 95, 167, 169
PUSHBUTTON SWITCHES 188
PUSHBUTTONS 190

Q

Q-PHASE DETECTOR 211, 212
QUAD BILATERAL SWITCH 157
QUAD OP-AMP IC 124
QUADRATURE 213
QUADRATURE OSCILLATOR 213, 239
QUESTION 22
QUIESCENT OPERATING POINT 5

R

RAG 60
RAIL 147
RANGE CAPACITOR 153
RANGE RESISTORS 51
RANGE SWITCH 51, 153
RATED ARMATURE VOLTAGE 181
RATIO 128
RC NETWORKS 138
RC4136 124
REACTANCE 91
READINGS 91, 98
READOUT 95
REAR CHANNEL 126
REAR CHANNEL AMPLIFIER 119, 120
REAR CHANNEL GAIN CONTROL 120

IC Design Projects

REAR CHANNEL INFORMATION 119
REAR CHANNEL POWER AMPLIFIER 123
REAR CHANNEL SIGNAL 119
REAR CHANNEL SPEAKERS 120
REAR GAIN CONTROL 123
REAR POWER AMPLIFIER 120
REAR SPEAKERS 119, 126
REAR WALL 126
REAR-CHANNEL AMPLITUDE 122
REAR-CHANNEL GAIN CONTROL 123
RECEIVER 217, 219, 222-224
RECEIVER CIRCUIT 219, 222
RECESSED ELEMENT 134
RECHARGEABLE BATTERY 52
RECORDING 127, 137
RECTANGULAR WAVES 167
RECTIFIER 5, 7, 12, 60, 239
RECTIFIER BRIDGE 189
RECTIFIER CIRCUIT 7, 8
RECTIFIER DIODE 13, 54
RECTIFIER OUTPUT 14
RED 102, 163
REFERENCE CLOCK 92
REFERENCE FREQUENCY 145
REFERENCE OUTPUT TERMINAL 147
REFERENCE SIGNAL 143, 145
REFERENCE SIGNAL FREQUENCY 143
REFERENCE VOLTAGE 16, 27, 31, 33, 36, 44,
   47, 96, 98, 99, 115, 168
REFERENCE VOLTAGE CIRCUIT 33
REFERENCE VOLTAGE SOURCE 15
REFERENCE VOLTAGE TERMINAL 169
REFERENCE VOLTAGE ZENER DIODE 36, 38
REGULATED OUTPUT VOLTAGE 38
REGULATION FACTOR 7, 14
REGULATOR 5, 12, 14, 28, 31-34, 36, 43, 45, 47,
   55, 67
REGULATOR DESIGNS 47
REGULATOR IC 91
REGULATOR POWER DISSIPATION 34
RELAY 197, 198, 206, 208, 219, 222-224
RELAY COIL 205, 206, 208
RELAY COIL IMPEDANCE 197
RELAY CONTACTS 197
RELAY LOAD SWITCHING 205
RELAY OUTPUT 219
RELIABLE SERVICE 208
REMOTE CONTROL 225
REMOTE CONTROL PROJECT 217
REMOTE CONTROL SYSTEM 217, 219, 224
REMOTE CONTROL TRANSMITTER 184
REMOTE SENSE TERMINALS 57, 63
REPETITIVE PEAK (RPI) 13
RESET 98, 168
RESET CIRCUITS 95
RESET SIGNAL 169, 195
RESET SWITCH 99, 102, 103
RESET TERMINAL 173

RESET TRANSISTOR 169
RESETTING 167
RESIDENTIAL ENVIRONMENTS 119
RESISTANCE 34, 75, 131, 184
RESISTANCES 131
RESISTIVE DIVIDER 31
RESISTOR 16, 31, 36, 47, 51, 59, 68, 69,
   74, 77, 94, 98, 99, 101, 102, 107-109,
   113, 114, 133-135, 138, 139, 146, 147,
   150, 153, 155, 157, 158, 162, 167, 168,
   170, 171, 181, 183, 184, 187-189, 191, 195,
   197, 202-205, 218, 220, 221, 223
RESISTOR CHAIN 188
RESISTOR NETWORK 50, 51, 52
RESISTOR VALUES 51, 79, 173
RESPONSE SPEED 146
REST RECOVERY CYCLE 218
REVERSE CURRENT GAIN 74
REVERSE VOLTAGE 65
REVERSE-BIASED 74
REVERSE-BIASED ISOLATION JUNCTIONS 112
RHEOSTAT 181, 239
RIAA EQUALIZATION 137, 138
RIAA PREAMPLIFIER 137, 139
RIAA PREAMPLIFIER STAGE 138
RIGHT CHANNEL 124
RIGHT CHANNEL SIGNAL 120
RIGHT INPUT SIGNAL 120
RIPPLE 12, 59
RIPPLE COMPONENT 8, 67, 222
RIPPLE DC VOLTAGE 124, 159
RIPPLE EFFECT 17
RIPPLE FACTOR 10, 12, 14
RIPPLE FREQUENCY 8, 10, 12
RIPPLE REJECTION FACTOR 36
RIPPLE VOLTAGE 8
RMS VOLTAGE 214
RMS VOLTS 7
ROLL-OFF SLOPES 132, 136
ROOM 195
ROTARY SWITCH 49
RPI 13
RPI CURRENT FLOW 13

S

SAFE LIGHT 205, 206, 208
SAFE-AREA COMPENSATION 32
SAFE-AREA PROTECTION CIRCUIT 34
SALLEN 115
SALLEN AND KEY 132
SAMPLED FUNCTION 145
SAMPLING VOLTAGES 33
SATELLITE 217
SATURATED CIRCUITRY 79
SATURATION 74, 76, 113, 115
SATURATION MODE 76, 77
SCALP 129, 134

Index

SCHEMATIC 49, 54, 57, 67, 91, 98, 121, 131, 137, 139, 157, 161, 162, 173, 175, 181, 205, 211, 213, 218, 219, 222
SCHOTTKY DIODE 80
SCHOTTKY TRANSISTOR 80
SCHOTTKY TTL GATES 80
SCRAP 159, 163, 223
SCRATCH 27
SCREEN 20
SCREEN DISPLAY 22
SCREEN OUTPUT OPTION 22
SECOND HAND 199
SECONDARY 239
SECONDARY RATING 189
SECONDARY TERMINALS 6
SECONDARY VOLTAGE 7
SECONDARY WINDING 6, 8, 10
SECONDS 173
SECURITY SYSTEM 157, 201
SELF-OSCILLATING SWITCHING REGULATOR CIRCUIT 31
SEMICONDUCTOR ESSENTIALS 8
SEMIDISPOSABLE ELECTRODES 134
SENSE DIODES 61
SENSE LEAD CONNECTIONS 61
SENSE LEADS 60
SENSE LINES 59, 60
SENSING SURFACE 134
SENSING WIRES 60
SEQUENTIAL OPERATION 195
SEQUENTIAL TIMING CIRCUITS 167
SERIES 86
SERIES CHOKE 14
SERIES COMBINATION 76
SERIES INDUCTOR 14
SERIES OPERATION 65
SERIES PASS ELEMENT 29
SERIES-CONNECTED RHEOSTAT 181
SERIES-PASS ELEMENTS 27, 28
SERIES-PASS REGULATOR 59
SERIES-PASS TRANSISTORS 30, 36, 38, 39, 59
SET 98, 168
SEVEN-SEGMENT DECODER 95
SEWING NEEDLE 102, 163
SHARPS 185
SHIRT POCKET 218
SHOCK HAZARDS 129, 135, 136
SHORT 161
SHORT CIRCUIT 57, 59
SHORT CIRCUITS 101, 122, 125, 140, 159, 160, 163, 184, 198, 203, 207, 223
SHORT MODE 206
SHORT POSITION 205, 208
SHORT-CIRCUIT PROTECTION 31, 112
SHORT-CIRCUIT PROTECTION CIRCUITS 188
SHORT-CIRCUITED LOAD 112
SHORT-CIRCUITED OUTPUT 31
SHORT-TERM VOLTAGES 134
SHORT/LONG SWITCH 205
SHORTS 155
SHUNT 12
SHUNT CAPACITORS 14
SHUNT RECTIFIER 132
SIDE WALLS 126
SIDEBANDS 145
SIGNAL 54, 99, 108, 109, 111, 114, 115, 125, 126, 128, 129, 131, 133, 137, 139, 143, 144, 173, 212, 217, 219, 220
SIGNAL BYPASS CAPACITORS 67
SIGNAL CONFIGURATION 75
SIGNAL FREQUENCY 145
SIGNAL GENERATOR 135
SIGNAL OUTPUT 219
SIGNAL SOURCE 120, 129, 139
SIGNAL VOLTAGES 129
SILICON 83
SILICON DIE 28
SILICON PLANAR 83
SILICON TRANSISTOR 28
SILVER PLATING 134
SILVER-SILVER CHLORIDE 134
SILVER-SILVER CHLORIDE PELLET ELECTRODE 134
SINE WAVE 126, 153
SINGLE WINDING SECONDARY 49
SINGLE-ENDED POWER SUPPLIES 20, 147
SINK 169
SINK CURRENT 94
SINUSOIDAL 239
SINUSOIDAL WAVEFORM 143
SIREN 157-160, 224
SKIN 135
SKIN POTENTIALS 133
SKIN SURFACE 129
SKIN SURFACE ELECTRODES 134
SLEW RATE 111, 239
SOLDER 60
SOLDER BRIDGES 55, 61, 101, 125, 140, 155, 159, 160, 163, 184, 190, 198, 203, 207, 223
SOLDER CONNECTION 163
SOLDER HEAT DAMAGE 163
SOLDER JOINTS 55
SOLDER SHORTS 52
SOLDERING 140, 155, 159, 163, 184, 198, 203, 207, 223
SOLDERING GUN 60
SOLDERING IRON 60, 89, 101, 159, 163
SOLDERING IRON HEAT 133
SOLDERING TECHNIQUES 101
SOUND 126, 137, 162, 201
SOUND EFFECTS 120, 121, 193
SOUND PATTERN 130
SOUND-EFFECTS GENERATORS 157
SOUNDTRACK 157
SOURCE 83, 84, 169
SOURCE CURRENT 83

SOURCE FOLLOWER 133, 149, 150
SOURCE IMPEDANCE 127, 134
SOURCE-TO-SUBSTRATE JUNCTION 84
SPACE TELEMETRY 143
SPEAKER 119, 120, 123, 126, 133, 202, 204
SPEAKER CABINET 163
SPEAKER COIL 202
SPEAKER ENCLOSURE 163
SPECIFICATION 112
SPEECH 120
SPEED 181
SPEED CONTROL 181, 183
SPIKE VOLTAGES 197
SPONGE 60
SPST PUSHBUTTON SWITCHES 190
SPURIOUS SIGNALS 128, 129
SQUARE WAVE 95, 114, 115, 153, 156, 167, 204
SQUARE WAVE INPUT SIGNAL 149
SQUARE WAVE OUTPUT 102, 156
SQUARE-WAVE DIFFERENTIAL-MODE INPUT SIGNAL 111
SQUARE-WAVE OUTPUT WAVEFORM 115
SSI (SMALL SCALE INTEGRATION) 73, 79, 89
STABLE CONTROLLER 167
STABLE STATE 171
STANDARD VALUE RESISTOR 189, 190
STATIC CHARGES 159
STATIC DAMAGE 159
STATIC ELECTRICITY 88
STEEL 190
STEP-DOWN TRANSFORMER 7, 49, 187, 240
STEP-UP TRANSFORMER 7, 240
STEREO CHANNELS 119
STEREO DEMODULATOR 144
STEREO PREAMPLIFIER 137, 139
STEREO SYSTEM 119
STEREO SYSTEMS 120
STEREO TELEVISION 119, 120
STEREO VIDEO CASSETTE RECORDER 119
STORAGE TIME 80
STRAY CAPACITANCE 93
SUBASSEMBLY 60
SUBAUDIO FREQUENCIES 120
SUBSTRATE 83, 84, 88
SUBSTRATE POTENTIAL 88
SUBWAYS 181
SUBWOOFER 121
SUBWOOFER SIMULATOR 120, 121, 122
SUMMING AMPLIFIER 111, 113, 120, 189, 191
SUMMING ATTENUATOR 189
SUPPLY CURRENT 77, 110
SUPPLY VOLTAGE 88, 133, 170, 173, 205
SUPPORT COMPONENTS 98, 101
SURFACE SKIN VOLTAGES 133
SURROUND SOUND 119, 126
SURROUND-SOUND DECODER 119, 124, 125
SURROUND-SOUND SYSTEM 119

SWITCH 31, 92, 93, 95, 98, 99, 101, 103, 120, 132, 133, 135, 139, 153, 155-157, 169, 188, 189, 192, 205-208, 218
SWITCH CONTROL 157
SWITCH TRANSISTOR 31
SWITCH-MODE INTEGRATED CIRCUIT REGULATOR 28
SWITCH-MODE POWER SUPPLY 27
SWITCH-MODE REGULATORS 29
SWITCHABLE 205
SWITCHED WAVEFORM 31
SWITCHING CIRCUIT 31
SWITCHING FUNCTIONS 137
SWITCHING MODE 87, 88
SWITCHING POINT 85
SWITCHING REGULATOR 27, 29-31
SWITCHING REGULATOR CIRCUIT 42
SWITCHING SPEEDS 79, 80, 85, 87, 91
SWITCHING SWEEP 87
SWITCHING THRESHOLD VOLTAGE 85
SYMMETRIC SQUARE WAVE OUTPUT 149
SYMMETRICAL BIPOLAR POWER SUPPLIES 53, 55
SYNCHRONIZER CIRCUITS 95
SYSTEM 83, 220, 224, 225

T

TAG STRIPS 51
TANK CIRCUIT 92
TANTALUM CAPACITOR 203
TAPE DECK 137
TAPE RECORDER 139
TAPE SOURCE 139
TDA-2002 188
TDA2002 123
TELEMETRY RECEIVERS 147
TELEVISION 120
TEMPERATURE 110, 167
TEMPERATURE DRIFT 28
TEMPERATURE RATINGS 89
TERMINAL VOLTAGES 49
TERMINALS 44, 61, 62, 63, 65, 84, 95, 109, 212
TEST CIRCUIT 102, 162, 164
TEST CIRCUIT COMPONENT VALUES 163
TEST CIRCUIT POWER SUPPLY 164
TEST LEAD 199, 204
TEST POINT 61, 103, 161
TEST POINT VOLTAGES 61
TEST SIGNAL 126
TESTING 52, 55, 61, 125
THERMAL PROTECTION CIRCUITS 47
THERMAL SHUTDOWN 32
THERMOCOUPLE 133
THETA 129, 135
THETA BRAIN WAVES 127
THIEF 201
THIEVES 195
THRESHOLD 168, 169

Index

THRESHOLD CONTROL 131
THRESHOLD INPUTS 169, 170, 171, 188
THRESHOLD LEVELS 173
THRESHOLD VOLTAGE 175
THROWBACK TRANSIENT CURRENTS 30
THUNDER 121
THYRISTORS 181
TIL906-1 219
TIME 91, 151, 167, 195
TIME CAPACITOR 218
TIME CONSTANT 240
TIME DELAY 167, 175, 176
TIME INTERVAL 175, 176, 220
TIME PERIOD DATA 193
TIME PERIODS 185
TIME-DELAY GENERATION CIRCUITS 167
TIMER 176, 205, 207, 208, 218
TIMER IC 167
TIMER PACKAGE 167
TIMER POSITION 205, 208
TIMER/OFF/FOCUS SWITCH 205
TIMES 173
TIMING 198, 200
TIMING CAPACITOR 92, 169, 170, 173, 181, 188, 189, 214
TIMING CIRCUITS 167
TIMING COMPONENTS 169, 171, 220
TIMING CYCLE 169, 173, 176
TIMING DELAY 205
TIMING DIAGRAM 196
TIMING INTERVAL 173, 205
TIMING RESISTOR 188, 189, 214, 218
TIMING RESISTOR CHAIN 190
TIMING RESISTOR TERMINAL 213
TO-220 PLASTIC PACKAGE 33
TO-3 METAL PACKAGE 33
TO-39 PACKAGE 47
TONE 130, 131, 133, 135, 162, 164, 204
TONE CENTER FREQUENCY 164
TONE CONTROL 139
TONE CONTROL CIRCUIT 137, 139
TONE DECODER 147, 211-214
TONE DECODER IC 212
TONE DECODER STAGE 220
TONE GENERATOR 130, 131
TONE-CONTROL STAGE 139
TONE-DECODER LM567 ICS 220
TOTEM-POLE CIRCUIT 76
TOTEM-POLE OUTPUT CIRCUIT 75
TOTEM-POLE OUTPUT STAGE 75
TOUCH-TONE DECODING 213
TOYS 83
TRACTION CONTROL 181
TRAILING WAVEFORMS 167
TRANSDUCER 133
TRANSFER CHARACTERISTIC 78, 85
TRANSFER CHARACTERISTICS 85

TRANSFORMER 5-8, 10, 22, 49, 51, 54, 57, 60, 67, 68, 91, 94, 124, 137, 147, 154, 155, 159, 187, 197, 205, 207, 222
TRANSFORMER CURRENT RATING 22
TRANSFORMER SECONDARY VOLTAGE 7, 49, 55
TRANSFORMER SECONDARY WINDING 12
TRANSFORMER SPECIFICATIONS 7
TRANSFORMER-RATED POWER LOAD 7
TRANSIENT 144
TRANSIENT RESPONSE 33, 54, 68, 91, 137, 205, 223
TRANSIENT SUPPRESSOR 221
TRANSISTOR 16, 27, 28, 31, 34, 49, 50-52, 59-61, 73-80, 83, 86, 107, 108, 110, 131, 133, 158, 160, 162, 163, 169, 170, 183, 197, 198, 202, 203, 218, 219, 222, 223
TRANSISTOR AMPLIFIER CIRCUITS 107
TRANSISTOR BETA 17
TRANSISTOR CHOPPER 181
TRANSISTOR COMMON-EMITTER AMPLIFIER 220
TRANSISTOR INVERTER SWITCHES 196
TRANSISTOR SERIES REGULATOR 15, 16
TRANSISTOR SOCKET 184
TRANSISTOR SWITCH 197, 198, 211
TRANSISTOR SWITCHES 197
TRANSISTOR TIME CONSTANTS 79
TRANSISTOR TURNOFF TIME 80
TRANSISTOR-TRANSISTOR LOGIC (TTL) 57, 63, 73, 91
TRANSMITTER 217-220, 223
TRANSMITTER CHANNEL 220
TRANSMITTER CIRCUIT 217
TRANSMITTER FREQUENCIES 223
TRANSMITTER SIGNAL 224
TREBLE BOOST 137, 139
TREBLE CONTROLS 137
TREBLE CUT 139
TREBLE TONE CONTROL CIRCUIT 139
TREMOLO 185, 189, 193
TREMOLO CIRCUIT 187-189
TREMOLO CONTROL 189
TRIANGULAR WAVE 153, 156
TRIANGULAR WAVE OUTPUT 156
TRIANGULAR WAVEFORM 115
TRIGGER 168, 188
TRIGGER INPUT 169, 170, 171, 173
TRIGGER LEVEL 99
TRIGGER PULSE 115, 173, 205
TRIGGERING 167
TRIMMER POTENTIOMETER 188, 189, 190, 192, 199, 220
TRISTATE LOGIC SWITCH 93, 94
TRISTATE OUTPUT DEVICES 89
TRUTH TABLES 78, 86
TTL 73, 97, 99
TTL CIRCUITS 57, 63, 91, 167
TTL DEVICES 87
TTL GATE 73-79, 87-89

Page 259

TTL ICS 73, 79
TTL INTEGRATED CIRCUIT 78
TTL LOGIC FAMILY 79
TTL NAND GATE 77
TTL SPECIFICATIONS 78
TUBE AMPLIFIERS 107
TUBE OPERATIONAL AMPLIFIER 107
TUNE 190
TUNED CIRCUITS 147
TURNTABLE RUMBLE 138
TWO-CHANNEL INFRARED REMOTE CONTROL 217
TWO-CHANNEL SYSTEM 217
TWO-FREQUENCY OSCILLATOR 213
TWO-OUTPUT VOLTAGES 67
TWO-PHASE 213
TWO-STATE BINARY CIRCUIT 168
TWO-TONE SOUND 201

U

ULTRASONIC CONTROLS 213
UNITY 122
UNITY GAIN 98, 111, 113, 121, 122, 123, 131
UNITY GAIN SOURCE FOLLOWER 133
UNITY GAIN SUMMING AMPLIFIER 189
UNITY-GAIN BANDWIDTH 111
UNIVERSAL HEAT SINK 60
UNREGULATED INPUT SUPPLY 30
UNREGULATED INPUT VOLTAGE 28, 33, 34
UPPER FREQUENCY LIMIT 148
USER 220

V

VA TRANSFORMER 7
VALUES 77, 102, 171, 223
VARIABLE ELEMENT 22
VARIABLE GAIN CIRCUITS 139
VARIABLE GAIN STAGE 132, 136, 137, 139
VARIABLE RESISTANCE 28
VARIABLE RESISTOR 34, 181
VARIABLE-CURRENT LIMITING CIRCUIT 42
VARYING DC VOLTAGE 5
VAS (VOLT-AMPERES) 7
VCO 143-151, 153, 157, 158, 162, 211, 212, 214
VCO FREQUENCY 144, 146, 148, 150, 151
VCO FREQUENCY OF OSCILLATION 149
VCO INPUT 162
VCO INPUT TERMINAL 157
VCO OPERATING FREQUENCY 157
VCO OUTPUT 162
VCR 119, 137, 139
VIDEO CIRCUITS 111
VISUAL DISPLAY 161
VLSI 89
VLSI (VERY LARGE SCALE INTEGRATION) 83
VLSI TECHNOLOGY 83
VOICE CHANNEL 120
VOLT REFERENCE 44

VOLTAGE 5, 12, 31, 33, 34, 36, 38, 43, 47, 53, 52, 67, 84, 88, 89, 91, 108, 110, 127, 129, 133, 138, 150, 157, 159, 188, 212
VOLTAGE CONVERSION 31
VOLTAGE DIFFERENCE 28, 36, 47
VOLTAGE DIVIDER 133
VOLTAGE DROP 31, 36, 50, 52, 57, 59, 60, 158
VOLTAGE FLUCTUATIONS 111
VOLTAGE FOLLOWER 16, 120
VOLTAGE GAIN 108, 188, 220
VOLTAGE LEVEL 8, 162
VOLTAGE RATING 49, 68
VOLTAGE REFERENCE AMPLIFIER 36
VOLTAGE REFERENCES 27
VOLTAGE REGULATOR 34, 36, 40, 43, 47, 54, 55, 57, 68, 116, 137, 183, 187, 205, 222
VOLTAGE REGULATOR CIRCUIT 27, 40
VOLTAGE SELECTION 49
VOLTAGE SOURCE 5, 59, 108, 127, 129, 150
VOLTAGE WAVEFORMS 153
VOLTAGE-CONTROLLED OSCILLATOR (VCO) 143
VOLTAGE-FOLLOWER AMPLIFIER 113
VOLTMETER 61, 63, 156
VOLTMETER NEEDLE 156
VOLTS 5, 12, 212
VOLUME 135
VOLUME CONTROL 124, 131, 137, 188

W

WAIL 158
WARBLE ALARM 201, 203, 204
WARBLE OSCILLATOR 204
WARNING SOUND 201
WATCHES 83
WATTS 7
WAVEFORMS 8, 10, 31, 79, 114, 131, 153, 155, 163, 170, 173, 175, 183, 202
WHITE 163
WIDE-BAND FSK DEMODULATION 213
WINDINGS 155, 181
WIRE 60, 102, 133, 163, 164
WIRE-WRAPPED 133
WIRED-AND LOGIC 89
WIRED-OR LOGIC 89
WIREWRAP 94
WIRING 61, 62, 208
WORKING VOLTAGE 12
WRISTWATCH 199

Y

YELLOW 163
YIELDS 28

Z

ZENER 16
ZENER CURRENT 17

Index

ZENER DIODE 14-17, 36, 38, 61, 148, 150
ZENER DIODE REFERENCE VOLTAGE 16
ZENER DIODE REGULATOR 14
ZENER DIODE SHUNT REGULATOR 14
ZENER VOLTAGE 15, 17
ZERO FREQUENCY 162
ZERO LEVEL 85
ZERO POTENTIOMETER 95
ZERO VOLTS 5
ZERO-TRIMMER POTENTIOMETER 93

Howard W. Sams
A Bell Atlantic Company

Your Technology Connection to the Future!

Now You Can Visit Howard W. Sams & Company <u>On-Line</u>:
http://www.hwsams.com

Gain Easy Access to:

- The **PROMPT Publications** catalog, for information on our *Latest Book Releases*.
- The **PHOTOFACT Annual Index**.
- Information on Howard W. Sams' Latest Products.
- *AND MORE!*

PROMPT®
PUBLICATIONS

CALL 1-800-428-7267 TODAY FOR THE NAME OF YOUR NEAREST PROMPT PUBLICATIONS DISTRIBUTOR